Forgotten Chemistry

A Refresher Course

Johanna Holm
Writer and Editor
American Council on Education

All inquiries should be addressed to:
Barron's Educational Series, Inc.
250 Wireless Boulevard
Hauppauge, New York 11788
www.barronseduc.com

Library of Congress Control Number: 2006013340

ISBN-13: 978-0-7641-3317-6
ISBN-10: 0-7641-3317-9

Library of Congress Cataloging-in-Publication Data

Holm, Johanna.
Forgotten chemistry : a refresher course / Johanna Holm.
p. cm.
Includes bibliographical references and index.
ISBN-13: 978-0-7641-3317-6 (alk. paper)
ISBN-10: 0-7641-3317-9 (alk. paper)
1. Chemistry, Organic. I. Title.

QD251.3.H65 2006
547—dc22

2006013340

PRINTED IN THE UNITED STATES OF AMERICA

9 8 7 6 5 4 3 2 1

CONTENTS

Preface

Forgotten Chemistry is a user-friendly guide to the major topics of first-year college chemistry. These ideas in simplified form are also the core of the high school chemistry program. The approach used in *Forgotten Chemistry* is suitable for those who have truly forgotten chemistry and for those who need to refresh their earlier coursework.

Forgotten Chemistry provides you with the tools and skills you will need for your journey into chemistry. The presentation of concepts and strategies in small, easier to manage packets offers you an opportunity to learn, practice, and then build on these skills. Each example is laid out in full detail so that the example itself answers your questions as you examine it. The examples are structured to cover all variations of the topic at hand; so it is advisable to analyze each one. Every topic discussed is accompanied by questions and answers designed to strengthen your understanding and improve your confidence.

Unit 15 of *Forgotten Chemistry* addresses the math and calculator skills you will need for success in chemistry. Often a lack of success in chemistry is more a math problem than a chemistry problem. If this applies to you, then working through Unit 15 first would be an excellent beginning. However you approach using *Forgotten Chemistry,* it is hoped that it will set you firmly on whatever path it is that you have in mind.

UNIT 1

The Building Blocks of Chemistry

Atoms, Elements, Compounds, and Molecules

Success in chemistry has been linked to the ability to see things "in your mind's eye." As you work your way through this self-instruction course, you will find activities designed to allow you to do this. Just as the letters of the alphabet can be used to form every word in our language, the elements of the periodic table can be used to form every substance in the world. Atoms of these elements join together to form the molecules and compounds that make up foods, furniture, automobiles, people, trees—all the matter that there is. In this unit, you will examine the ideas behind elements, atoms, molecules, and compounds. By the end of the unit you will have a mental picture of each of these ideas that will help you navigate through the concepts of chemistry.

In the periodic table, each square represents one element. All the diverse items of our world—healthy foods, deadly poisons, automobiles, books, and drugs, to name just a few—are made from combinations of these elements. Some of the elements are present in our world in large quantities, whereas others are scarce.

Each element has a story of its own, and these stories are part of the fabric of history and literature. The Mad Hatter in Lewis Carroll's *Alice in Wonderland* was likely mad because of the mercury used in the hat-making industry. As scientists began to realize that there must be a connection between hatters' working conditions and the unusual incidence of mental illness among hatters, mercury came to the fore as an environmental hazard. Lead has had a similar toxic past and has even been implicated in the decline of the Roman Empire. Ancient Romans were fond of boiling wine in lead bowls in order to produce syrup that was used as a sweetener. As science evolved, it became known that part of the sweetness was caused by a lead compound formed during this process.

Symptoms of lead poisoning include tiredness and lack of mental abilities, which could certainly have interfered with Roman leadership and might have caused the fall of the Roman Empire.

ATOMS

A single atom is composed of one nucleus and one cloud of negative electrons surrounding the nucleus. The nucleus is made up of protons (positive particles) and neutrons (neutral particles). The atoms of each element have a number of protons unique to that element. For example, each atom of the element chlorine has 17 protons. The cloud of electrons that surrounds the nucleus has a very tiny mass. This mass is so small that we say it has no mass at all. This cloud of electrons will be examined thoroughly in the unit on atomic structure. For now, just visualize it as having the same number of electrons as the number of protons in the atom.

In a pure gold ring, there are a large number of gold atoms but only one element, gold. In distinguishing between the terms *atom* and *element*, element is not specific about the number, whereas atom refers to just one of the smallest parts of an element. The behavior of each element is dictated by the internal structure of the atoms of that element.

The primary subatomic particles in each atom are protons, electrons, and neutrons. The mass (weight) and charge of each are given in Table 1.1.

TABLE 1.1: MASS AND CHARGE OF SUBATOMIC PARTICLES

Subatomic Particle	Charge	Mass (Atomic Mass Units)
Proton	+1	1
Electron	−1	0
Neutron	0	1

Mass Number and Atomic Number

The mass of an atom is one of its important characteristics. Notice in Table 1.1 that only the proton and the neutron have mass. It is for this reason that their sum is called the mass number of the element. **The mass number is the total number of protons and neutrons in a single atom.**

The number of protons in an atom is called the atomic number of the element. It is the proton number that makes the element what it is. For example, calcium has an atomic number of 20; therefore, each atom of calcium has 20 protons. An atom that does not have 20 protons cannot be calcium. Except in very unusual cases, the number of protons does not change. The alchemists of long ago, who tried to make gold out of other elements, did not realize that their efforts were doomed by this problem. To make gold, they would have had to add protons to create atoms with gold's 79 proton count.

Atoms are neutral because the number of positive protons is equal to the number of negative electrons.

Using the Periodic Table to Find Atomic Number and Mass Number

The periodic table (Figure 1.1) has been called "the chemist's cheat sheet" because of the amount of information it provides. Look at the square that has the letters Ca in it. The first letter for an element is always capitalized, and the second letter, if there is one, is lowercase. Notice that there are two numbers in the square. The smaller of the two is the number of protons (the atomic number). It is also equal to the number of electrons. The larger number in each square in the periodic table is the mass number of the element (the sum of the protons and neutrons). For our purposes, this mass number should be rounded to the nearest whole number. For example, the larger number in the block for Br (bromine) is 79.9, and we will round this off to 80.

In the periodic table (Figure 1.1) there is enough information to tell the number of protons, electrons, and neutrons in every element. Sometimes it is easier to use an alphabetized list of the elements rather than searching for them in the periodic table (Figure 1.2).

Example:

Gold (Au)

The smaller number is 79; so there are 79 protons in each gold atom.

The larger number is 197; so to find the number of neutrons, subtract 79 from 197 to get 118.

The number of electrons is the same as the number of protons; so there are 79 electrons in each gold atom.

EXERCISES

Use the periodic table (Figure 1.1) and the list of elements and their symbols (Figure 1.2) to answer the following questions.

1. Find the number of protons, electrons, and neutrons for the element silicon (Si), a major component of sand.

   Protons
   Electrons
   Neutrons

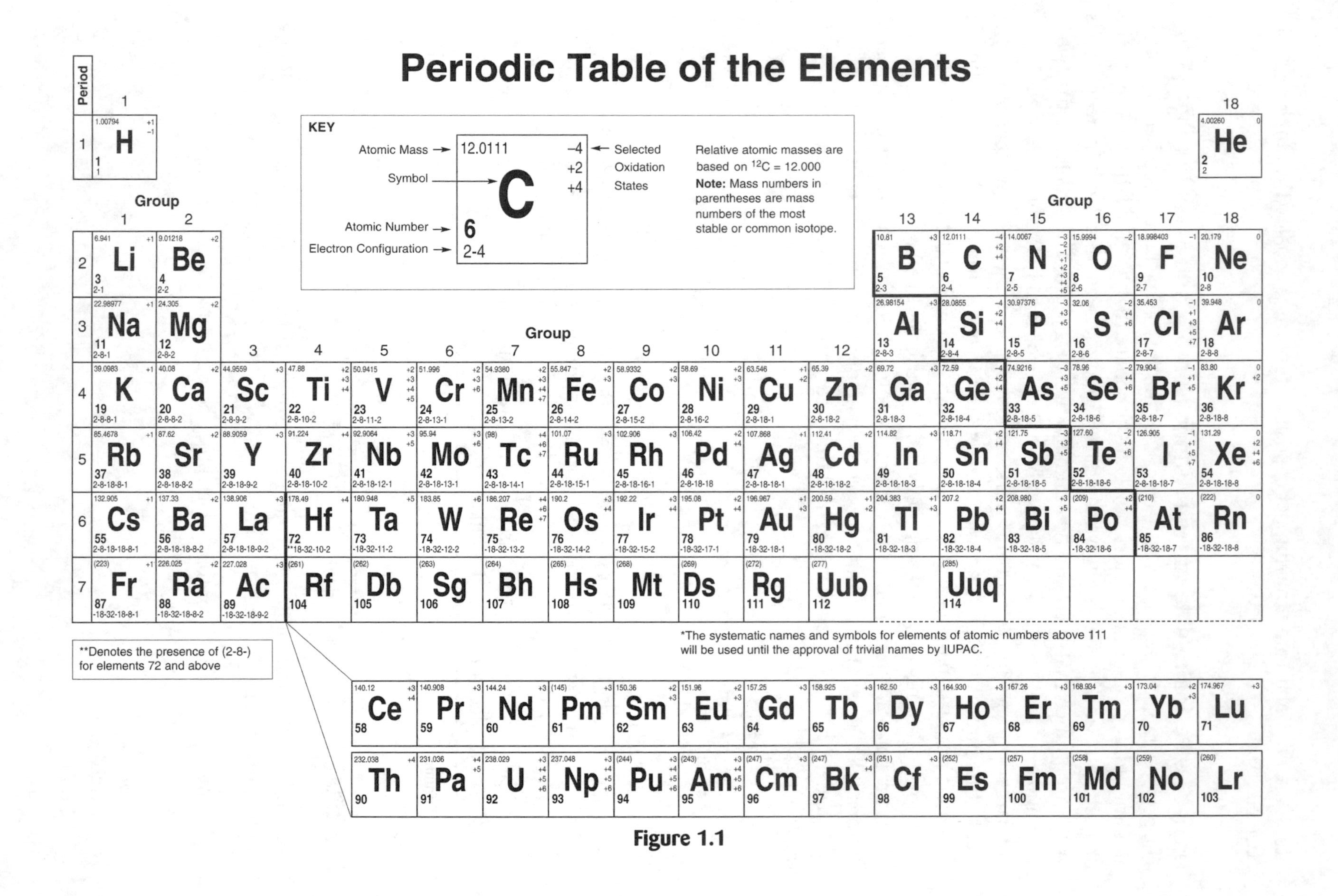
Periodic Table of the Elements
KEY
Atomic Mass
Symbol
Atomic Number
Electron Configuration
12.0111
C
6
2-4
-4
+2
+4
Selected Oxidation States
Relative atomic masses are based on ¹²C = 12.000
Note: Mass numbers in parentheses are mass numbers of the most stable or common isotope.
Period
Group
*The systematic names and symbols for elements of atomic numbers above 111 will be used until the approval of trivial names by IUPAC.
**Denotes the presence of (2-8-) for elements 72 and above

Figure 1.1

THE CHEMICAL ELEMENTS

(Atomic masses in this table are based on the atomic mass of carbon-12 being exactly 12.)

Name	*Symbol*	*Number*	*Mass*
Actinium	Ac	89	(227)
Aluminum	Al	13	26.98
Americium	Am	95	(243)
Antimony	Sb	51	121.75
Argon	Ar	18	39.95
Arsenic	As	33	74.92
Astatine	At	85	(210)
Barium	Ba	56	137.34
Berkelium	Bk	97	(247)
Beryllium	Be	4	9.01
Bismuth	Bi	83	208.98
Bohrium		107	(262)
Boron	B	5	10.81
Bromine	Br	35	79.90
Cadmium	Cd	48	112.40
Caesium	Cs	55	132.91
Calcium	Ca	20	40.08
Californium	Cf	98	(251)
Carbon	C	6	12.01
Cerium	Ce	58	140.12
Chlorine	Cl	17	35.45
Chromium	Cr	24	52.00
Cobalt	Co	27	58.93
Copper	Cu	29	63.55
Curium	Cm	96	(247)
Darmstadtium	Ds	110	(269)
Dabnium		105	(262)
Dysprosium	Dy	66	162.50
Einsteinium	Es	99	(254)
Erbium	Er	68	167.26
Europium	Eu	63	151.96
Fermium	Fm	100	(257)
Fluorine	F	9	19.00
Francium	Fr	87	(223)
Gadolinium	Gd	64	157.25
Gallium	Ga	31	69.72
Germanium	Ge	32	72.59
Gold	Au	79	196.97
Hafnium	Hf	72	178.49
Hassium	Hs	108	(269)
Helium	He	2	4.00
Holmium	Ho	67	164.93
Hydrogen	H	1	1.008
Indium	In	49	114.82
Iodine	I	53	126.90
Iridium	Ir	77	192.2
Iron	Fe	26	55.85
Krypton	Kr	36	83.80
Lanthanum	La	57	138.91
Lawrencium	Lr	103	(262)
Lead	Pb	82	207.19
Lithium	Li	3	6.94
Lutetium	Lu	71	174.97
Magnesium	Mg	12	24.31
Manganese	Mn	25	54.94
Meitnerium	Mt	109	(268)
Mendelevium	Md	101	(258)

Name	*Symbol*	*Atomic Number*	*Atomic Mass*
Mercury	Hg	80	200.59
Molybdenum	Mo	42	95.94
Neilsbohrium	Ns	107	(262)
Neodymium	Nd	60	144.24
Neon	Ne	10	20.18
Neptunium	Np	93	237.05
Nickel	Ni	28	58.71
Niobium	Nb	41	92.90
Nitrogen	N	7	14.01
Nobelium	No	102	(259)
Osmium	Os	76	190.2
Oxygen	O	8	16.00
Palladium	Pd	46	106.4
Phosphorus	P	15	30.97
Platinum	Pt	78	195.09
Plutonium	Pu	94	(244)
Polonium	Po	84	(210)
Potassium	K	19	39.10
Praseodymium	Pr	59	140.90
Promethium	Pm	61	(145)
Protactinium	Pa	91	231.04
Radium	Ra	88	(226)
Radon	Rn	86	(222)
Rhenium	Re	75	186.2
Rhodium	Rh	45	102.91
Roentgenium	Rg	111	(272)
Rubidium	Rb	37	85.47
Ruthenium	Ru	44	101.07
Rutherfordium	Rf	104	(261)
Samarium	Sm	62	150.35
Scandium	Sc	21	44.95
Seaborgium	Sg	106	(263)
Selenium	Se	34	78.96
Silicon	Si	14	28.09
Silver	Ag	47	107.89
Sodium	Na	11	22.99
Strontium	Sr	38	87.62
Sulfur	S	16	32.06
Tantalum	Ta	73	180.95
Technetium	Tc	43	(99)
Tellurium	Te	52	127.60
Terbium	Tb	65	158.92
Thallium	Tl	81	204.37
Thorium	Th	90	232.03
Thulium	Tm	69	168.93
Tin	Sn	50	118.69
Titanium	Ti	22	47.90
Tungsten	W	74	183.85
Uranium	U	92	238.03
Vanadium	V	23	50.94
Xenon	Xe	54	131.30
Ytterbium	Yb	70	173.04
Yttrium	Y	39	88.91
Zinc	Zn	30	65.37
Zirconium	Zr	40	91.22

*A number in parentheses is the mass number of the most stable isotope.

Figure 1.2

2. Find the number of protons, electrons, and neutrons for the element argon (Ar), found in most lightbulbs.

   Protons
   Electrons
   Neutrons

3. What element has 17 protons?

4. What element has 6 neutrons?

5. What element has 8 electrons?

6. How many elements have 28 protons?

7. What is the atomic number of magnesium (Mg)?

8. What is the atomic mass of uranium (U)?

Answers

1. protons 14
   electrons 14
   neutrons 14
2. protons 18
   electrons 18
   neutrons 22
3. chlorine
4. carbon and boron
5. oxygen
6. one
7. 12
8. 238

MOLECULES

A molecule is a combination of two or more atoms that are joined or bonded together. These atoms can be of the same element or of different elements.

A most familiar molecule is water. H_2O is the chemical formula for water and a shorthand way to say that each water molecule is made up of two hydrogen atoms and one oxygen atom as shown in Figure 1.3. In the figure, the lines between hydrogen and oxygen represent bonds.

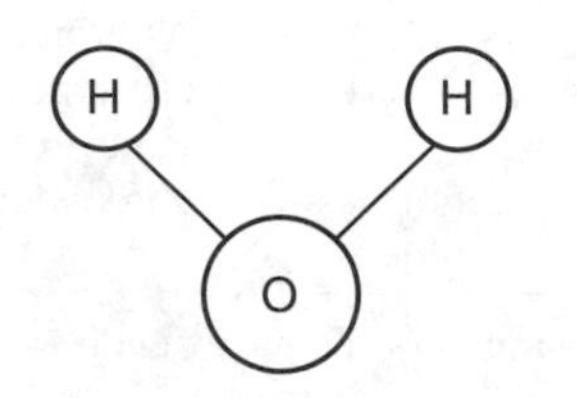

Figure 1.3: Water molecule (H_2O)

Subscripts are small numbers placed after the symbol for an element to show how many atoms of that element are present in the molecule. For example, H_3PO_4 is phosphoric acid and is found in most carbonated beverages. The formula shows that one molecule of phosphoric acid has three atoms of hydrogen, one atom of phosphorus, and four atoms of oxygen.

The following are examples of molecules.

N_2 (nitrogen)	H_2O (water)	O_3 (ozone)
H_2SO_4 (battery acid)	$NaHCO_3$ (baking soda)	

Notice that the common factor among these molecules is that there are at least two atoms in each one. The elements can be the same, or they can be different. Notice also that the number of elements does not have a particular pattern.

COMPOUNDS

A compound is made of at least two atoms that must be of different elements. The word *molecule* applies to more substances than the word *compound*. For example, oxygen (O_2) is a molecule but not a compound. It is a molecule because it has two atoms. It cannot qualify as a compound because it has only one element, oxygen.

EXERCISES

Answer the following questions about molecules, compounds, elements, and atoms.

1. Battery acid is H_2SO_4 (sulfuric acid). How many elements are present?

   How many atoms?
   Is it a compound?
   Is it a molecule?

2. Which of these are molecules? Ca, HNO_3, F_2, CO

3. In the following diagram the different shadings in the circles represent different elements.

   How many compounds are there?
   How many molecules?
   How many atoms?
   How many elements?

4. What is the minimum number of atoms in a molecule?
5. What is the minimum number of atoms in a compound?
   What is the minimum number of elements in a compound?
6. Do three atoms of oxygen bonded together form a molecule?
   A compound?
   An element?
7. Hydrogen peroxide, found in many medicine cabinets, has the formula H_2O_2.
   How many elements are in this formula?

   How many atoms?
   Is it a compound?
   Is it a molecule?
8. Find Fe in the periodic table. Which of the terms atom, element, compound, and molecule can be applied to Fe?

Answers

1. Three elements: hydrogen (H), oxygen (O), and sulfur (S). There are seven atoms present (2 of H + 1 of S + 4 of O = 7). It is a compound because there is more than one element present. It is a molecule because there are more than two atoms joined together.
2. The molecules are HNO_3 (nitric acid), CO (carbon monoxide), and F_2 (fluorine). All have one or more atoms, and it is not necessary that they be different elements.
3. There are 2 compounds (the circles are joined and are of different elements). There are 4 molecules (joined). There are 11 atoms (circles) and 3 elements (black, white, and striped).
4. There must be at least two atoms in a molecule.
5. There must be at least two atoms in a compound. There must also be at least two elements in a compound.
6. Three atoms of oxygen bonded together would make a molecule but not a compound. It is an element.
7. There are two elements: hydrogen and oxygen. Hydrogen peroxide has four atoms: two of oxygen and two of hydrogen. It is a molecule, having at least two atoms, and it is also a compound, having at least two elements.
8. Fe is the symbol for iron. The terms *atom* and *element* apply.

IONS

An *ion* is an atom or a molecule that has either gained or lost electrons. Some elements are electron losers, whereas others are electron gainers. Generally, elements on the left side of the periodic table are electron losers, and those on the right are electron gainers.

Positive Ions

A positive ion is formed when an atom loses one or more electrons. Because electrons have a negative charge, and the protons (with a positive charge) now outnumber the electrons, the resulting ion has a positive charge. The size of the charge depends on the number of electrons lost. You will always know the number of electrons involved. For example, calcium always loses two electrons to form a +2 ion. This calcium ion is shown as Ca^{2+}. The element potassium (K) loses just one electron to form a +1 ion and is shown as K^+.

The following shows the formation of a sodium +1 ion from a sodium atom. A sodium atom has 11 protons, 11 electrons, and 12 neutrons (a shown for Na in the periodic table). To become a +1 ion, the sodium atom has to lose 1 electron:

Na	+11	−11	12 neutrons
Na^+	+11	−10	12 neutrons

Look at the charges on the ion. There are 11 pluses and 10 minuses, for a net result of +1: Na^+.

Our taste buds are sodium ion detectors. The neutral sodium atom (Na) does not taste salty. In fact, tasting a sodium atom would be risky business, as sodium atoms react violently with water. Whereas atoms and ions of the same element have some features in common, they are also very different.

To sum up: When atoms become ions, only the number of electrons changes. The numbers of protons and neutrons do not change.

Here is another example of a positive ion. Aluminum forms +3 ions. The process looks like this.

Al	+13	−13	14 neutrons
Al^{3+}	+13	−10	14 neutrons

Aluminum foil and aluminum cans represent the neutral atom (Al) form of the element. The ionic form (Al^{3+}) looks like table salt and is found in some deodorants and plant fertilizers.

If you see the symbol for an element and there is no charge, that means that it is not an ion but rather is the neutral atom of the element, having the same numbers of protons and electrons.

EXERCISES

You will need the periodic table and the element/symbol list to answer the following questions about positive ions.

1. Look up the element silver (Ag) as found in silver jewelry. It has

   +_______ (protons) and –_______ (electrons). How would these numbers change for the silver +1 ion used in photographic film?

   +_______ (protons) and –________ (electrons).

2. The calcium that should be included in your diet is the +2 ion. In the spaces below, write the number of pluses (protons) and minuses (electrons) in both the neutral calcium atom and the +2 calcium ion.

   Ca: _________ protons, _______ electrons

   Ca^{2+}: ________ protons, _______ electrons

3. What happens to the number of neutrons when a neutral atom becomes a positive ion?

4. Look back at question 2. Would another way to make a calcium +2 ion be to add two pluses (protons) to the calcium atom? Explain.

5. Fill in the number of protons, electrons, and neutrons for Mg and Mg^{2+}.

Mg		Mg^{2+}
______________	protons	______________
______________	electrons	______________
______________	neutrons	______________

6. Which of the following is a suitable route for the neutral atom of potassium (K) to become the potassium ion (K^+)?
   (A) gaining one proton
   (B) gaining one electron
   (C) losing one proton
   (D) losing one electron

7. The barium in barium sulfate ($BaSO_4$) is the Ba^{2+} ion. Does this ion have a different number of neutrons than the neutral barium atom?

8. The lithium used in some pharmaceutical compounds for treating depression is the +1 ion. How many protons does this ion have? How many electrons?

9. In general, how does a neutral atom become a positive ion?

Answers

1. Ag: 47 protons, 47 electrons

   Ag^{+}: 47 protons, 46 electrons

2. Ca: 20 protons, 20 electrons

   Ca^{2+}: 20 protons, 18 electrons

3. Nothing. Neutrons are not involved in ion formation.

4. No. If the number of protons is changed, the element will be changed and will no longer be calcium.

5.

Mg		Mg^{2+}
12	protons	12
12	electrons	10
12	neutrons	12

6. (D) Losing one electron is the only way a neutral atom can become a positive ion. Gaining a proton would not work, as that would change the identity of the element itself.

7. The barium ion and the barium atom have the same number of neutrons.

8. The lithium ion has 3 protons and 2 electrons, giving it a +1 charge.

9. For any neutral atom to become a +1 ion, it must lose one electron.

Negative Ions

Negative ions are formed when a neutral atom gains one or more electrons. Once more, neutrons are not involved in the process. The chorine atom becomes a chloride ion (Cl^{-}) by the addition of one electron, as shown here:

Cl	17 protons	17 electrons	So there is one more minus.
Cl^{1-}	17 protons	18 electrons	

The chloride ion is familiar to us as part of table salt (NaCl) and is used in water purification.

The atoms of some elements gain not one but two electrons. Oxygen is one such element:

O	8 protons	8 electrons	So there are two more minuses.
O^{2-}	8 protons	10 electrons	

EXERCISES

You will need the periodic table to answer the following questions about negative ions.

1. The fluoride ion is added to most public drinking water supplies in order to reduce dental decay. This ion carries a –1 charge. Write the proton/electron numbers for both the fluoride ion and the fluorine neutral atom.
2. Because negative ions feature an excess of negative charges, can negative ions be made by subtracting pluses (protons)? Explain.
3. In the –3 nitrogen ion, how many protons are there? Electrons? Neutrons?
4. The sulfur ion is a –2 ion. How did it get to be an ion starting from the neutral sulfur atom?
5. How does the X^{2-} ion differ from the neutral X atom?
6. Most people buy table salt that is iodized. Iodized means that iodide ions (I^-) have been added to promote thyroid health. How many protons does this iodide ion have? Electrons? Neutrons?
7. With respect to the P^{3-} ion, which of the following is true?
   (A) protons > electrons
   (B) protons < electrons
   (C) protons = electrons
8. How does the selenium atom (Se) become the selenium –2 ion?

Answers

1. F: 9 protons, 9 electrons
   F^{1-}: 9 protons, 10 electrons
2. No. The number of protons cannot change without changing the element itself.

3. 7 protons, 10 electrons, 7 neutrons
4. By adding two electrons.
5. Every –2 ion differs from its neutral atom by having two more electrons.
6. The iodide ion has 53 protons, 54 electrons, and 74 neutrons.
7. (B) For an ion to have a negative charge, the negative electrons have to outnumber the positive protons.
8. The neutral selenium atom gains two electrons to become the selenium ion with a –2 charge.

Making Compounds Out of Ions

Now you will use these building blocks to create different compounds. You have likely heard it said that opposites attract. The same is true for ions. Positive ions and negative ions come together to make compounds.

There are rules for this compound formation:

- The positive ion comes first in the formula.
- The number of positive charges must be made to equal the number of negative charges.
- Subscripts are added wherever necessary in order to have equal numbers of positive and negative charges.

To make a compound out of H^+ and O^{2-}, you need two H^+ ions; so two pluses are added to the two minuses of the O^{2-} ion. We all recognize the result as H_2O.

In order to place the proper subscripts in a formula, follow this plan. Take the number next to the + and make it the subscript for the negative ion. Take the number next to the – and make it the subscript of the positive ion. If the number is a 1, it is understood and is not written. That is, H_2O is not written as H_2O_1.

Examples

Making compounds out of ion combinations

Al^+ and Cl^-:	Al_1Cl_3	$AlCl_3$
Na^+ and S^{2-}:	Na_2S_1	Na_2S
Fe^{3+} and O^{2-}:	Fe_2O_3	
Mg^{2+} and O^{2-}:	Mg_2O_2	MgO*

*Subscripts must be reduced to simplest terms. This will be explained further in the next chapter.

Remember that the object is for the number of pluses to be equal to the number of minuses.

EXERCISES

Create neutral compounds from each of the following combinations of positive and negative ions.

1. Fe^{3+} and Cl^{-}
2. Mg^{2+} and S^{2-}
3. Na^{+} and O^{2-}
4. Al^{3+} and O^{2-}
5. Ba^{2+} and Br^{1-}
6. K^{1+} and I^{1-}
7. Ca^{2+} and O^{2-}
8. Fe^{3+} and S^{2-}
9. Ba^{2+} and N^{3-}
10. Li^{+} and S^{2-}

Answers

1. $FeCl_3$
2. MgS
3. Na_2O
4. Al_2O_3
5. $BaBr_2$
6. KI
7. CaO
8. Fe_2S_3
9. Ba_3N_2
10. Li_2S

UNIT 2

Chemical Nomenclature

Naming Compounds and Writing Their Formulas

The word *nomenclature* simply means naming. You have already seen some chemical formulas such as NaCl, H_2O, and H_2SO_4. There are many times when having names for chemicals is more convenient than having their formulas. Chemical naming falls into two main groups: that for ionic compounds formed of positive and negative ions, and that for nonionic compounds.

IONIC COMPOUNDS

You had some experience with ionic compounds in Unit 1. For example, you were able to write the correct formula for the compound formed from a calcium +2 ion and a chloride –1 ion by using the idea that there must be the same number of positive and negative charges. This led you to write the correct formula by adding two chloride ions to one calcium ion. The positive ion always goes first; so the formula is $CaCl_2$. The nomenclature (naming) of ionic compounds involves three kinds of compounds: binary compounds, compounds of polyatomic ions (ions with more than one atom), and compounds of multivalent ions (ions with more than one charge).

Binary Compounds

A binary compound is formed of only two elements. A typical binary compound is table salt (NaCl). The naming strategy for binary compounds is to write the name of the first element, add part of the name of the second element, and end with -ide. The part of the name of the second element used is best illustrated by examples. If the second element is chlorine, change it to chloride. If the second element is sulfur, change it to sulfide. Oxygen is oxide, fluorine is fluoride, and so on (Table 2.1).

TABLE 2.1: SIMPLE IONS

Ion	Name	Ion	Name
H^+	hydrogen	H^-	hydride
Li^+	lithium	F^-	fluoride
Na^+	sodium	Cl^-	chloride
K^+	potassium	Br^-	bromide
Mg^{2+}	magnesium	I^-	iodide
Ca^{2+}	calcium	O^{2-}	oxide
Ba^{2+}	barium	S^{2-}	sulfide
Al^{3+}	aluminum	N^{3-}	nitride
Ag^+	silver		

Table 2.2 shows examples of writing formulas and names. Remember to insert subscripts to make the positive charges equal the negative charges.

TABLE 2.2: WRITING FORMULAS AND NAMES

Elements Involved	Ions	Formula	Name
Sodium and fluorine	Na^+ F^-	NaF	sodium fluoride
Aluminum and chlorine	Al^{3+} Cl^-	$AlCl_3$	aluminum chloride
Potassium and oxygen	K^+ O^{2-}	K_2O	potassium oxide
Calcium and sulfur	Ca^{2+} S^{2-}	CaS*	calcium sulfide

*Because calcium and sulfur have the same number of charges (2), no subscript is necessary as the charges are already equal.

EXERCISES

Complete the following table.

Elements Involved	Ions	Formula	Name
1. Lithium and bromine			
2. Magnesium and iodine			
3. Sodium and oxygen			
4. Silver and sulfur			
5. Aluminum and oxygen			
6. Magnesium and sulfur			
7. Calcium and oxygen			
8. Barium and chlorine			
9. Potassium and sulfur			

Answers

	Ions		Formula	Name
1.	Li^+	Br^-	$LiBr$	lithium bromide
2.	Mg^{2+}	I^-	MgI_2	magnesium iodide
3.	Na^+	O^{2-}	Na_2O	sodium oxide
4.	Ag^+	S^{2-}	Ag_2S	silver sulfide
5.	Al^{3+}	O^{2-}	Al_2O_3	aluminum oxide
6.	Mg^{2+}	S^{2-}	MgS	magnesium sulfide (reduce to lowest terms)
7.	Ca^{2+}	O^{2-}	CaO	calcium oxide (reduce to lowest terms)
8.	Ba^{2+}	Cl^-	$BaCl_2$	barium chloride
9.	K^+	S^{2-}	K_2S	potassium sulfide

Compounds with Polyatomic Ions

Look at Table 2.3. **A polyatomic ion is a group of two or more elements that stay together and act as though they are one ion.** Locate the sulfate ion in the table. What is its formula? How many elements are in it? How many oxygen atoms are there? What is the sulfate ion's charge?

TABLE 2.3: POLYATOMIC IONS

Ion	Name	Ion	Name
NH_4^+	ammonium	CO_3^{2-}	carbonate
NO_2^-	nitrite	ClO^-	hypochlorite
NO_3^-	nitrate	ClO_2^-	chlorite
SO_3^{2-}	sulfite	ClO_3^-	chlorate
SO_4^{2-}	sulfate	CrO_4^{2-}	chromate
OH^-	hydroxide	MnO_4^-	permanganate
CN^-	cyanide	$C_2H_3O_2^-$	acetate
PO_4^{3-}	phosphate	O_2^{2-}	peroxide

The sulfite ion is written as SO_4. It has two elements, four oxygen atoms, and a charge of –2.

When you write the formula for a compound involving a polyatomic ion, you go about it in the same general way as for a binary compound. There are two differences. The name of the polyatomic ion is used, and the polyatomic ion needs to have parentheses around it. Study the following examples of polyatomic ions to see how this works.

Note: Pay close attention to the formulas of polyatomic ions, as some differ only slightly. For example, the sulfate (SO_4) and sulfite (SO_3) ions differ by only one letter in the name and by only one oxygen atom in the formula.

Sodium and nitrate	Na^+ NO_3^-	$NaNO_3$	sodium nitrate
	Because there are one plus and one minus, nothing else needs to be done.		
Calcium and nitrate	Ca^{2+} NO_3^-	$Ca(NO_3)_2$	calcium nitrate
	Look at the parentheses. If they were not present, the reader might think there are 32 oxygen atoms. The parentheses show that the whole nitrate ion is included in the doubling process.		
Ammonium sulfate	NH_4^+ SO_4^{2-}	$(NH_4)_2SO_4$	ammonium sulfate
	Notice that parentheses are needed around the ammonium ion so it will not look like there are 42 Hs. Parentheses are not needed around the sulfate as there is no ambiguity.		

EXERCISES

Use Tables 2.1 and 2.3 for this practice.

A. Write the correct formula for these chemical names.

1. sodium sulfate
2. barium nitrate
3. aluminum nitrite
4. potassium carbonate
5. ammonium phosphate
6. magnesium sulfite
7. calcium hydroxide
8. lithium nitrate
9. ammonium sulfate
10. sodium carbonate

B. Write the correct name for each of these formulas.

1. K_3PO_4
2. $MgSO_4$
3. $Al(OH)_3$
4. Li_2SO_3
5. $CaCO_3$
6. $NaOH$
7. NH_4OH
8. $Ca_3(PO_4)_2$
9. $BaSO_3$
10. $Al_2(SO_4)_3$

Answers

A.

1. Na_2SO_4
2. $Ba(NO_3)_2$
3. $Al(NO_2)_3$
4. K_2CO_3
5. $(NH_4)_3PO_4$
6. $MgSO_3$
7. $Ca(OH)_2$
8. $LiNO_3$
9. $(NH_4)_2SO_4$
10. Na_2CO_3

B.

1. potassium phosphate
2. magnesium sulfate
3. aluminum hydroxide
4. lithium sulfite
5. calcium carbonate
6. sodium hydroxide
7. ammonium hydroxide
8. calcium phosphate
9. barium sulfite
10. aluminum sulfate

Compounds with Multivalent Ions

Inspect the listing of multivalent ions in Table 2.4 and note that some ions have more than one possible charge. **A multivalent ion is one that can have more than one charge, such as Fe^{3+} and Fe^{2+}.** When writing formulas and names for compounds having one of these ions, it is necessary to show which ion is involved.

TABLE 2.4: MULTIVALENT IONS

Ion	Systematic Name	Older Name
Fe^{2+}	iron(II)	ferrous
Fe^{3+}	iron(III)	ferric
Cu^{+}	copper(I)	cuprous
Cu^{2+}	copper(II)	cupric
Co^{2+}	cobalt(II)	cobaltous
Co^{3+}	cobalt(III)	cobaltic
Sn^{2+}	tin(II)	stannous
Sn^{4+}	tin(IV)	stannic
Pb^{2+}	lead(II)	plumbous
Pb^{4+}	lead(IV)	plumbic

When you combine the iron ion and the chloride ion, the possibilities are

Fe^{+2} and $Cl^{-} = FeCl_2$ and Fe^{3+} and $Cl^{-} = FeCl_3$

A naming system using Roman numerals specifies the charge of the multivalent ion:

$FeCl_2$ = iron(II) chloride	and	$FeCl_3$ = iron(III) chloride

The only time that Roman numerals are added to names is to distinguish between multivalent ions. Here are two more examples:

tin(IV) fluoride	Sn^{4+} and F^-	SnF_4
lead(II) nitrate	Pb^{2+} and NO_3^-	$Pb(NO_3)_2$

EXERCISES

Remember to use Table 2.4 in answering the following questions.

A. Write the chemical formulas for the following compounds:

1. iron(II) hydroxide
2. copper(I) sulfate
3. cobalt(II) chloride
4. lead(IV) oxide
5. iron(III) sulfate
6. lead(II) nitrate
7. tin(II) hydroxide
8. copper(I) phosphate
9. tin(IV) fluoride
10. iron(III) oxide

B. Write the names that describe the following formulas:

1. Fe_2O_3
2. $CuSO_3$
3. $SnCl_2$
4. $Co_2(SO_4)_3$
5. PbO_2
6. $CoSO_4$
7. $Sn(NO_3)_2$
8. $Fe(ClO_3)_3$
9. $Co(NO_3)_3$
10. $Sn(CrO_4)_2$

Answers

A.

1. $Fe(OH)_2$ The roman numeral II shows that the Fe is Fe^{2+}; so you will need two –1 hydroxide ions.

2. Cu_2SO_4 The copper has a +1 charge; so you will need two of them to go with the –2 sulfate ion.

3. $CoCl_2$ The cobalt has a charge of +2; so you need two –1 chloride ions.

4. PbO_2 Because lead has a charge of +4 and oxygen has a charge of –2, you need two oxygen atoms to go with the one lead atom. If your answer was Pb_2O_4, remember to reduce the subscripts to lowest terms.
5. $Fe_2(SO_4)_3$ Here, use of the crossover technique is helpful. The +3 from the Fe becomes the subscript for the sulfate.
6. $Pb(NO_3)_2$ This formula results in +2 from the lead and –2 from two of the –1 nitrate ions.
7. $Sn(OH)_2$ The reasoning here is the same as in question 6.
8. Cu_3PO_4 Because PO_4 has a –3 charge, you will need +3 from the copper ion.
9. SnF_4 The formula here has +4 and –4.
10. Fe_2O_3 Once again, the crossover technique is a quick way to write a formula with the same number of pluses as minuses.

B.

1. iron(III) oxide Oxygen is the unvarying element; so you can count on its being –2. With three oxygen atoms in the formula, the negative part is –6. Now look at the iron; in order for two iron atoms to equal a charge of +6, each must have a charge of +3.
2. copper(II) sulfite Use the same strategy as for question 5. Look at the sulfite first because it does not vary. Its charge of –2 means that the copper must have a charge of +2; so copper(II) is correct.
3. tin(II) chloride Because there are two chloride ions, the total negative charge is –2, requiring +2 from the tin ion.
4. cobalt(III) sulfate As you examine the –2 charge of the sulfate ion, you will see that the total negative charge is –6. To offset the –6 a charge of +6 is needed. With two cobalt ions in the formula, each has a charge of +3.
5. lead(IV) oxide The reasoning for the rest of the answers follows the same principle requiring the same number of positive charges as negative ones.
6. cobalt(II) sulfate
7. tin(II) nitrate
8. iron(III) chlorate
9. cobalt(III) nitrate
10. tin(IV) chromate

The Classic Naming System for Multivalent Compounds

Although the Roman numeral system is most commonly used for multivalent ions, there is an older system that remains in use enough to be mentioned here. Look at Table 2.4 again. Examine the older names in the column on the far right. You will notice two things. All the endings are either -ic or -ous, and the beginnings of the names of the elements are not what you would expect. Because you will always have access to this table, memorizing is not required.

Continue to examine Table 2.4 to find the common feature among all the ions ending with -ic and all the ions ending with -ous. The -ous ending is used to designate the ion with the lower of the two possible charges for that ion, whereas the -ic ending denotes the ion with the higher charge. Look at the two iron ions; the +2 ion has the lower charge, has the -ous ending, and is named ferrous. The ion with the higher charge is in the +3 state, has the -ic ending, and is named ferric. As you have figured out by now, in order to know which name to use, you must first find out the charge of the ion.

Examples

Naming compounds using the old system

1. **$FeCl_3$.** The Cl ion has a –1 charge as shown in Table 2.1. There are three of them, and so they contribute –3. Because the positive charges must equal the negative charges, the iron must be Fe^{3+}. The +3 is the higher of the two possible charges for iron. For this reason, the name is ferric chloride.
2. **$CuSO_4$.** SO_4 has a –2 charge as shown in Table 2.1. Cu must therefore have a +2 charge to balance out the charges. The +2 charge is the higher of copper's possible charges. For this reason, the name is cupric sulfate.

Note: In one of the examples the +2 ion is -ous because it has the lower of the two possible charges, and in the other +2 ion is -ic because it has the higher of the two possible charges.

The iron in over-the-counter iron supplements frequently is ferrous sulfate, whereas iron in the ferric state, ferric chloride, was an old (not very effective) poison ivy remedy. Stannous fluoride was an ingredient in toothpaste until it was replaced with sodium fluoride. The terms *plumbous* and *plumbic* for lead have historic roots in the word *plumber*, originally used to refer to someone who works with lead pipes.

EXERCISES

Use the older, classic naming system in answering these questions.

A. Write the name for each of these compounds. Remember that Roman numerals are required.

1. PbO_2
2. Cu_2O
3. $FeCO_3$
4. SnF_2
5. Co_2S_3

B. Write the formula for each of these names.

1. plumbous nitrate
2. cupric fluoride
3. ferrous chloride
4. stannic oxide
5. cobaltous nitrate

Answers

A.

1. plumbic oxide
2. cuprous oxide
3. ferrous carbonate
4. stannous fluoride
5. cobaltic sulfide

B.

1. $Pb(NO_3)_2$
2. CuF_2
3. $FeCl_2$
4. SnO_2
5. $Co(NO_3)_2$

To review: Up to this point you have learned about the nomenclature of compounds formed from positive and negative ions of various kinds: single ions, polyatomic ions, and multivalent ions.

COMPOUNDS MADE OF NONMETALS

Look at the periodic table. There is a darker stairstep line dividing it into right and left portions. The elements to the left of this line are called metals, whereas those to the right are called nonmetals, with the exception of the column of elements on the right-hand border of the table. These elements in the last column are neither metals nor nonmetals but rather are called noble gases. Helium (He), the familiar gas used to fill up balloons, is the first of these noble gases.

Examples of compounds formed from nonmetals are

CO_2	N_2O_5	SO_3

In naming these compounds, you will need to use Table 2.5.

TABLE 2.5: PREFIXES FOR NAMING NONMETAL COMPOUNDS

Prefix	Number
mono-	1
di-	2
tri-	3
tetra-	4
penta-	5
hexa-	6
hepta-	7
octa-	8

Examine the names and formulas of these example compounds and notice the following.

- Prefixes are used before an element's name to indicate the number of these atoms in the formula.
- The prefix mono is used only if there is one atom of the second element in the compound. If there is just one atom of the first element, no prefix is used because the 1 is understood.
- The second element is named as if it were an ion, such as chloride, oxide, and so on.
  sulfur dioxide SO_2
  dinitrogen pentoxide N_2O_5
  carbon tetrachloride CCl_4

In working with formulas for nonmetals, notice that there is no balancing of charges as was the case for metal and nonmetal combinations. You will know the formula from its name, such as carbon dioxide. You will also be able to write the name if you know the formula.

EXERCISES

A. Write the chemical formula for each of these names.

1. sulfur trioxide
2. carbon monoxide
3. iodine trifluoride
4. carbon dioxide

B. Write the name for each of these chemical formulas.

1. N_2O
2. SF_6
3. CF_4
4. N_2O_3

Answers

A.

1. SO_3
2. CO
3. IF_3
4. CO_2

B.

1. dinitrogen monoxide
2. sulfur hexafluoride
3. carbon tetrafluoride
4. dinitrogen trioxide

Balancing Equations

Law of Conservation of Mass, Use of Coefficients, and Balancing Techniques

THE CHEMICAL EQUATION

A chemical equation is a statement of what is happening in a chemical reaction using chemical formulas. An example of a chemical equation is

$$HCl + NaOH \rightarrow NaCl + H_2O$$

The reactants are the substances on the left of the arrow in a chemical equation. They are the ingredients in the reaction. **The products are the substances on the right of the arrow in a chemical equation.** They are the result of the chemical reaction. The arrow in the chemical equation marks the transition from reactant to product. The arrow is read "yields" or "produces."

$$\text{Reactants} \xrightarrow{\text{yield}} \text{products}$$

The chemical equation of the reaction that takes place when charcoal briquettes burn in a grill is

$$C + O_2 \rightarrow CO_2$$

This equation states that charcoal (C) reacts with the oxygen in the air (O_2) to produce carbon dioxide (CO_2). The reactants are C and O_2, and the product is carbon dioxide (CO_2).

COUNTING ATOMS

To balance equations, it is necessary to be able to count the atoms represented. This counting strategy involves subscripts, parentheses, and coefficients.

Subscripts

From your previous formula work, you know that subscripts are the numbers that are written after the symbol for an element and placed slightly lower. In the formula for water, (H_2O) 2 is a subscript (O has a subscript of 1 that is understood). The subscript numbers indicate how many atoms of a particular element are in a formula. In the formula for battery acid (H_2SO_4), the subscripts show that there are two hydrogen atoms, one sulfur atom, and four oxygen atoms.

Parentheses

In chemistry parentheses are used to group elements together. For example, calcium nitrate has the formula $Ca(NO_3)_2$. The parentheses include the N and the O_3 in the doubling that the 2 indicates. The calcium nitrate formula has one calcium, two nitrogen, and six oxygen atoms. Notice that the calcium is not doubled because the parentheses do not include it. Calcium nitrate has a total of seven atoms in its formula.

Coefficients

A coefficient is a number placed in front of a formula to multiply the entire formula. For example, $3Ca(NO_3)_2$ indicates that there are three calcium nitrate formulas, and the 3 multiplies everything in the formula. The atom count looks like this:

Element	No. in the formula	×	Coefficient	=	Total atoms of element
Ca	1	x	3	=	3
N	2	x	3	=	6
O	6	x	3	=	18
Total atoms in $3Ca(NO_3)_2$				=	27

EXERCISES

For each of the following, determine the total number of atoms.

1. H_3PO_4
2. Fe_2O_3
6. $Al_2(SO_4)_3$
7. $2ZnCO_3$

3. $BaSO_3$
4. $Ca(OH)_2$
5. $(NH_4)_2S$
8. $3Al(NO_3)_3$
9. $4Fe(OH)_3$
10. $3NiF_2$

Answers

1. 3 H + 1 P + 4 O = 8 total atoms
2. 2 Fe + 3 O = 5 total atoms
3. 1 Ba + 1 S + 3 O = 5 total atoms
4. 1 Ca + 2 O + 2 H = 5 total atoms
5. 2 N + 8 H +1 S = 11 total atoms
6. 2 Al + 3 S +12 O = 17 total atoms
7. 2 Zn + 2 C + 6 O = 10 total atoms
8. 3 Al + 9 N + 27 O = 39 total atoms
9. 4 Fe + 12 O + 12 H = 28 total atoms
10. 3 Ni + 6 F = 9 total atoms

BALANCING CHEMICAL EQUATIONS

The law behind balancing equations is called the Law of Conservation of Mass. **The Law of Conservation of Mass says that mass is neither created nor destroyed in a chemical reaction.** In other words, the mass present on the reactant side of the equation must be equal to the mass on the product side of the reaction (Figure 3.1).

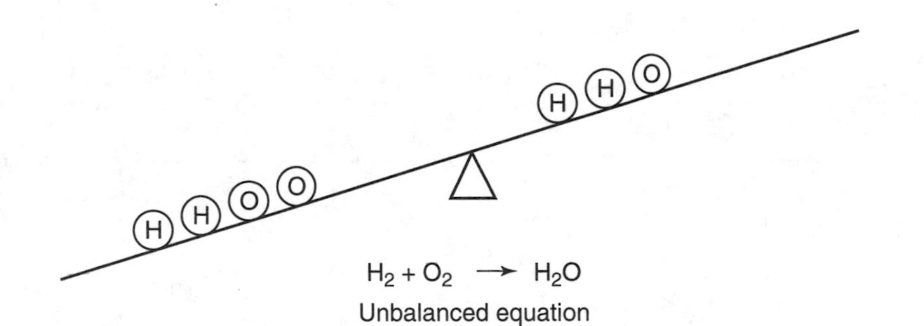

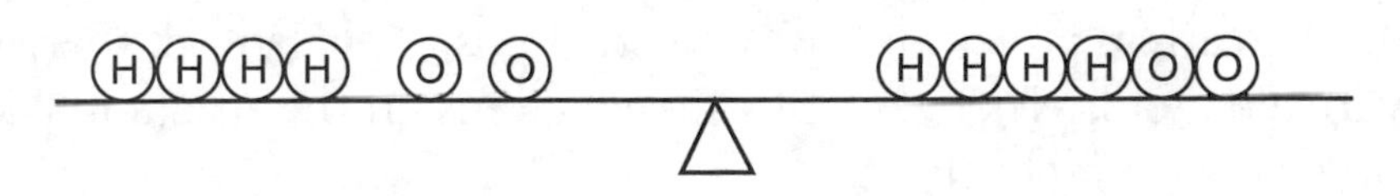

$2H_2 + O_2 \rightarrow 2H_2O$

Balanced equation

Figure 3.1: Unbalanced and balanced equations.

The balanced equation is the chemist's recipe. It tells what reactants to mix together, what the end product(s) will be, and the quantities of each. An example of a chemical equation (not yet balanced) is

$$CH_4 + O_2 \rightarrow CO_2 + H_2O$$

We will use this equation to address a variety of topics; so you will need to keep referring to it.

A word version of this equation states that methane (CH_4) plus oxygen (O_2) makes carbon dioxide (CO_2) and water (H_2O). This is the reaction that takes place when the burner of a gas stove is turned on.

EXERCISES

Look at the equation again and answer the following questions.

$$CH_4 + O_2 \rightarrow CO_2 + H_2O$$

1. How many carbon atoms are on the left? How many are on the right?
2. How many hydrogen atoms are on the left? How many are on the right?
3. How many oxygen atoms are on the left? How many are on the right?

Answers

1. 1, 1
2. 4, 2
3. 2, 2

Although the carbon atoms and the oxygen atoms are balanced, there is a problem with the hydrogen atoms. There are 4 hydrogen atoms on the reactant side, but only 2 hydrogen atoms on the product side.

You can tell that some numbers must be placed in front of the reactants and products to make the numbers on the left and right the same for each element. There are no rules for doing this, just some suggestions. One method is called balancing by inspection.

Balancing by Inspection

Look at the hydrogen atoms. There are 4 on the left and only 2 on the right. To correct this problem, you can put a 2 in front of the water molecule:

$$CH_4 \quad + \quad O_2 \quad \rightarrow \quad CO_2 \quad + \quad 2H_2O$$

Now the numbers of carbon atoms are equal on both sides (1) and the numbers of hydrogen atoms are equal on both sides (4), but there are only 2 oxygen atoms on the left and 4 oxygen atoms on the right. (Remember to count the oxygen atoms in both the CO_2 and the H_2O.)

Put a 2 in front of the O_2; now there are 4 oxygen atoms on the left, and the balanced equation is

$$CH_4 \quad + \quad 2O_2 \quad \rightarrow \quad CO_2 \quad + \quad 2H_2O$$

This is called a balanced equation because each element has the same number of atoms on the left side of the arrow as on the right side.

Reactant Side	Product Side
1 C	1 C
4 H	4 H
4 O	4 O (2 + 2)

What does the balanced equation tell you? It says that 1 molecule of CH_4 added to 2 O_2 molecules produces 1 molecule of CO_2 and 2 molecules of H_2O.

The order in which the reactants are written is not important. In the previous example you could have written $O_2 + CH_4$. The products could also have been written in any order, as long as they were on the right of the arrow. There can be any number of reactants and any number of products. For example, there might be one reactant and two products, three reactants and two products, and so on.

The following is another example of balancing an equation.

$$Zn \quad + \quad HCl \quad \rightarrow \quad H_2 \quad + \quad ZnCl_2$$

You can see that placing a coefficient of 2 in front of the HCl results in a balanced equation:

$$Zn \quad + \quad 2HCl \quad \rightarrow \quad H_2 \quad + \quad ZnCl_2$$

This balanced equation now has 1 zinc atom on each side of the equation, 2 hydrogen atoms on each side, and 2 chlorine atoms on each side. The coefficients (numbers placed in front of the reactant formulas and product formulas) must always be whole numbers, and if the number is a 1, it is said to be understood and is not actually shown. A slightly more complicated example is the equation for the burning of ethyl alcohol:

$$C_2H_5OH + \quad O_2 \quad \rightarrow \quad CO_2 \quad H_2O$$

First, when counting atoms, remember to consider the whole left side and the whole right side. In looking at just the hydrogen atoms on the left, how many are there? (6). How could you make the number of hydrogen atoms on the right equal 6? (By putting a 3 in front of the H_2O.)

Continuing to balance, the equation, how could you make the carbon atoms balance? (By placing a 2 in front of the CO_2.) At this point, the equation looks like this:

$$C_2H_5OH + \quad O_2 \quad \rightarrow \quad 2CO_2 \quad + \quad 3H_2O$$

We saved the oxygen for last because oxygen appears in so many places. This is always a good strategy. There are 7 oxygen atoms on the right; so there must be 7 on the left. There is 1 oxygen in the C_2H_5OH; so putting a 3 in front of the O_2, would result in 6 oxygen atoms there, giving 7 on the left. The balanced equation is

$$C_2H_5OH + \quad 3O_2 \quad \rightarrow \quad 2CO_2 \quad + \quad 3H_2O$$

Balancing equations is a "tinkering" or trial-and-error process. Using a pencil is a good idea. As you work at balancing equations, you will develop your own style.

EXERCISES

Balance the following equations by placing the correct coefficient in front of each formula. Remember that if that coefficient is 1, you do not write it.

1.	Cl_2	+	KBr	$\rightarrow$	Br_2	+	KCl
2.	NO			$\rightarrow$	N_2O	+	NO_2
3.	$Ba(NO_3)_2$	+	KF	$\rightarrow$	BaF_2	+	KNO_3
4.	HCl	+	$Ca(OH)_2$	$\rightarrow$	$CaCl_2$	+	H_2O

Answers

1. $Cl_2 \quad + \quad 2KBr \quad \rightarrow \quad Br_2 \quad + \quad 2KCl$

2. $3NO \quad \rightarrow \quad N_2O \quad + \quad NO_2$

3. $Ba(NO_3)_2 \quad + \quad 2KF \quad \rightarrow \quad BaF_2 \quad + \quad 2KNO_3$

4. $2HCl \quad + \quad Ca(OH)_2 \quad CaCl_2 \quad + \quad 2H_2O$

Balancing Equations by Common Multiple

Two other strategies are useful enough to be included here. One of them is similar to the idea of a least common multiple used in math. Suppose you have 2 oxygen atoms on the left of the equation and 3 oxygen atoms on the right? Think about both 2 and 3 fitting into the number 6 and use coefficients to make 6 oxygen atoms on both sides. An example of this is

Unbalanced equation:

$$KCl \quad + \quad O_2 \quad \rightarrow \quad KClO_3$$

Following the advice in the paragraph above, the 2 oxygens on one side of the equation and the 3 oxygens on the other side would both fit into the number 6. Keeping this 6 in mind, a coefficient of 3 on the reactant side and a coefficient of 2 on the product side would produce 6 oxygen atoms on each side of the equation.

$$KCl \quad + \quad 3O_2 \quad \rightarrow \quad 2KClO_3$$

The oxygen is now balanced, and one more coefficient (the 2 in front of the KCl) will balance the rest.

$$2KCl \quad + \quad 3O_2 \quad \rightarrow \quad 2KClO_3$$

EXERCISES

Balance the following equations by finding a common multiple.

1. $NH_4NO_2 \quad \rightarrow \quad N_2 \quad + \quad H_2O$

2. $Cr \quad + \quad O_2 \quad \rightarrow \quad Cr_2O_3$

3. $P_4 \quad + \quad H_2 \quad \rightarrow \quad PH_3$

4. $SO_2 \quad + \quad O_2 \quad \rightarrow \quad SO_3$

5. $Fe_2O_3 \quad + \quad C \quad \rightarrow \quad Fe_3O_4 \quad + \quad CO$

Answers

1. $NH_4NO_2 \rightarrow N_2 + 2H_2O$
2. $4Cr + 3O_2 \rightarrow 2Cr_2O_3$
3. $P_4 + 6H_2 \rightarrow 4PH_3$
4. $2SO_2 + O_2 \rightarrow 2SO_3$
5. $3Fe_2O_3 + C \rightarrow 2Fe_3O_4 + CO$

Temporary Use of Fractional Exponents

The last balancing equation strategy involves the temporary use of a coefficient that is not a whole number, such as ½. Notice the word *temporary*.

To explain, the following is the reaction that occurs when a butane lighter is lit.

$$C_4H_{10} + O_2 \rightarrow CO_2 + H_2O$$

As you work from left to right, the following coefficients are added.

$$C_4H_{10} + O_2 \rightarrow 4CO_2 + 5H_2O$$

When you add all the oxygen atoms on the right of the arrow (in both product molecules), you get 13 oxygen atoms. What can you put in front of the O_2 to come up with 13? There is no whole number that would work, but you could temporarily use 6½. At least the equation will be balanced, even though fractional numbers are allowed only on a temporary basis.

Remembering the idea of a temporary cofficient, you can double each of the coefficients and the equation will be balanced with all whole numbers. The final result will look like this:

$$2C_4H_{10} + 13O_2 \rightarrow 8CO_2 + 10H_2O$$

These are coefficients you would have had a hard time coming up with using the other systems.

EXERCISES

Balance these equations using fractional coefficients.

1. $C_6H_{14} + O_2 \rightarrow CO_2 + H_2O$
2. $O_2 + C_8H_{18} \rightarrow H_2O + CO_2$

3.	ZnS	+	O_2	→	ZnO	+	SO_2
4.	C_2H_6	+	O_2	→	CO_2	+	H_2O

Answers

1. temporary coefficients: 1, 9½, 6, 7; permanent coefficients: 2, 19, 12, 14
2. temporary coefficients: 1, 12½, 9, 8; permanent coefficients: 2, 25, 18, 16
3. temporary coefficients: 1, 1½, 1, 1; permanent coefficients: 2, 3, 2, 2
4. temporary coefficients: 1, 3½, 2, 3; permanent coefficients: 2, 7, 4, 6

EXERCISES

Use any system that works for you to balance the following equations by placing coefficients where appropriate.

1.	Co	+	O_2	→	Co_2O_3	(*Note:* Co is one element, cobalt.)	
2.	Al_2O_3			→	Al	+	O_2
3.	C_3H_8	+	O_2	→	H_2O	+	CO_2
4.	Al	+	HNO_3	→	H_2	+	$Al(NO_3)_3$
5.	PBr_3	+	H_2O	→	HBr	+	H_3PO_3
6.	NO	+	O_2	→	NO_2		
7.	CuO	+	H_2SO_4	→	H_2O	+	$CuSO_4$
8.	NH_3	+	Cl_2	→	NCl_3	+	NH_4Cl
9.	$AgNO_3$	+	Zn	→	Ag	+	$Zn(NO_3)_2$
10.	$NaClO_3$			→	$NaCl$	+	O_2
11.	C_6H_{14}	+	O_2	→	CO_2	+	H_2O
12.	Fe	+	O_2	→	Fe_2O_3		

Answers

1.	$4Co$	+	$3O_2$	$\rightarrow$	$2Co_2O_3$		
2.	$2Al_2O_3$			$\rightarrow$	$4Al$	+	$3O_2$
3.	C_3H_8	+	$5O_2$	$\rightarrow$	$4H_2O$	+	$3CO_2$
4.	$2Al$	+	$6HNO_3$	$\rightarrow$	$3H_2$	+	$2Al(NO_3)_3$
5.	PBr_3	+	$3H_2O$	$\rightarrow$	$3HBr$	+	H_3PO_3
6.	$2NO$	+	O_2	$\rightarrow$	$2NO_2$		
7.	The equation is balanced as is.						
8.	$4NH_3$	+	$3Cl_2$	$\rightarrow$	NCl_3	+	$3NH_4Cl$
9.	$2AgNO_3$	+	Zn	$\rightarrow$	$2Ag$	+	$Zn(NO_3)_2$
10.	$2NaClO_3$			$\rightarrow$	$2NaCl$	+	$3O_2$
11.	$2C_6H_{14}$	+	$19O_2$	$\rightarrow$	$12CO_2$	+	$14H_2O$
12.	$4Fe$	+	$3O_2$	$\rightarrow$	$2Fe_2O_3$		

Armed with the language of chemistry, you are now ready to continue your journey.

The Mole

The Mole Concept, Problems, and Applications

The balanced equation below is a recipe stating that for every 3 molecules of hydrogen (H_2) you need 1 molecule of nitrogen (N_2) to produce 2 molecules of ammonia (NH_3).

$$3H_2 \quad + \quad N_2 \quad \rightarrow \quad 2NH_3$$

In order for chemists to use balanced equations, they need a way to count molecules and atoms. As you can imagine, molecules and atoms are so tiny that it is impossible to count them out as you would apples and oranges.

Because such regular counting is impossible, chemists must have a different way to count. An analogous type of counting can be found in the banking industry. If you go to the bank with a huge number of pennies not wrapped in rolls, the bank staff can weigh them in order to compute their worth. Banks are able to do this because they know how much an average penny weighs.

The calculation looks like this:

$$\text{Weight of the pennies} \times \frac{\text{1 penny}}{\text{avg. wt. of 1 penny}} = \text{number of pennies}$$

A similar approach allows chemists to count. The first step in this process is to learn about an idea called molar mass.

MOLAR MASS

It is time for another look at the periodic table. Recall from Unit 1 that the larger of the two numbers in each square is called the atomic mass of the element. **Molar mass is the sum of the atomic masses of all the atoms in the formula of a compound or molecule.** Remember to round these atomic masses to the nearest whole number.

Examples

Calculations of molar mass. (We will deal with the units later.)

1. Calculate the molar mass of water (H_2O):

H	=	2 hydrogens	× 1 (atomic mass of H)	=	2
O	=	1 oxygen	× 16 (atomic mass of O)	=	<u>16</u>
		Molar mass of water		=	18

2. Calculate the molar mass of battery acid (H_2SO_4):

H	=	2 hydrogen	× 1 (atomic mass of H)	=	2
S	=	1 sulfur	× 32 (atomic mass of S)	=	32
O	=	4 oxygen	× 16 (atomic mass of O)	=	<u>64</u>
		Molar mass of battery acid		=	98

3. Calculate the molar mass of a component of rust [$Fe(OH)_3$]:

Fe	=	1 iron	× 56 (atomic mass of iron)	=	56
O	=	3 oxygen	× 16 (atomic mass of oxygen)	=	48
H	=	3 hydrogen	× 1 (atomic mass of hydrogen)	=	<u>3</u>
		Molar mass of $Fe(OH)_3$ =			107

Notice that the parentheses cause everything inside to be multiplied by the subscript that follows.

EXERCISES

Calculate the molar mass of each of the following.

1. $AgNO_3$
2. $Ca(OH)_2$
3. Zn
4. H_3PO_4
5. $(NH_4)_2S$

Answers

1. 170
2. 74
3. 65
4. 98
5. 68

COUNTING ATOMS AND MOLECULES

Now that you know how to calculate molar mass, what does it mean? As it turns out, every molar mass, expressed in grams, has the same number of molecules (or atoms if we are talking about a single element).

In other words, in the practice problems above, 170 grams of $AgNO_3$ has the same number of molecules as does 74 grams of $Ca(OH)_2$, as does 98 grams of H_3PO_4, and so on. In looking at the single element, zinc, we say that 65 grams of it has the same number of atoms because there are no molecules involved.

THE MOLE

This same number is called a mole. *Mole* is a counting word, just as *dozen* is a counting word. Although the word *dozen* makes everyone think of 12, the word *mole* makes chemists think of 6.02×10^{23}. **A mole is 6.02×10^{23}.** A mole of donuts is 6.02×10^{23} donuts. Two moles of marbles is $2 \times 6.02 \times 10^{23}$ marbles. The number 6.02×10^{23} is called Avogadro's number, after Amedeo Avogadro, the Italian scientist who was instrumental in its development. Now look at the molar masses that you calculated above:

Molar Mass	= Avogadro's Number	= 1 Mole
170 grams of $AgNO_3$	$= 6.02 \times 10^{23}$ molecules of $AgNO_3$	= 1 mole
74 grams of $Ca(OH)_2$	$= 6.02 \times 10^{23}$ molecules of $Ca(OH)_2$	= 1 mole
65 grams of Zn	$= 6.02 \times 10^{23}$ atoms of Zn	= 1 mole

From this mole–gram relationship come the units of molar mass, grams/mole. For example, the molar mass of silver nitrate ($AgNO_3$) is 170 grams/mole.

DIMENSIONAL ANALYSIS

The physical sciences use a problem-solving approach called dimensional analysis. Dimensional analysis requires conversion factors. A conversion factor is a numerator and a denominator that are equal to each other. Some conversion factors are

$$\frac{1 \text{ yard}}{3 \text{ feet}} \qquad \frac{3 \text{ feet}}{1 \text{ yard}} \qquad \frac{1 \text{ day}}{24 \text{ hours}} \qquad \frac{24 \text{ hours}}{1 \text{ day}} \qquad \frac{1 \text{ centimeter}}{10 \text{ millimeters}}$$

Notice that for every two units of measure that are equal to each other (such as 1 yard and 3 feet), two conversion factors can be written, one the reciprocal of the other. You will soon see how to decide which of the two to use.

Using Dimensional Analysis and Conversion Factors in Problem Solving

EXAMPLE

How many seconds are there in 1 week?

$$1 \text{ \cancel{week}} \times \frac{7 \text{ \cancel{days}}}{1 \text{ \cancel{week}}} \times \frac{24 \text{ \cancel{hours}}}{1 \text{ \cancel{day}}} \times \frac{60 \text{ \cancel{minutes}}}{1 \text{ \cancel{hour}}} \times \frac{60 \text{ seconds}}{1 \text{ \cancel{minute}}} = 604{,}800 \text{ seconds}$$

Examine the problem solution above and notice the following.

- The beginning of the problem is the amount given in the problem: 1 week.
- The beginning unit *week* means that the next conversion factor should have *week* in the denominator so *week* can cancel out.
- Because *days* appears in the numerator of the first factor, *days* must be in the denominator of the next factor.
- The factors continue on until the word *seconds* appears in the numerator and all the other words have canceled out.
- Each factor must be a true statement, for example, 1 day = 24 hours.

You probably could have determined how many seconds there are in 1 week without using dimensional analysis, but for more difficult problems this strategy can be a most valuable resource. Table 4.1 lists some standard measurements conversions.

Calculator Hints

There is a special calculator section in Unit 15. You may or may not need to refresh your calculator skills by using the information and practice in that section.

As a calculator strategy, the multiplication key is pressed before numbers appearing in the numerator, and the division key is pressed before numbers appearing in the denominator.

TABLE 4.1: TABLE OF MEASUREMENT CONVERSIONS

Length:			
	1 mile	=	5,280 feet
	1 kilometer	=	1,000 meters
	1 centimeter	=	10 millimeters
	1 meter	=	39.37 inches
	1 inch	=	2.54 centimeters
Volume:			
	1 gallon	=	4 quarts
	1 quart	=	0.95 liter
	1 quart	=	32 ounces
Mass:			
	1 kilogram	=	2.2 pounds
	454 grams	=	1 pound
	1 pound	=	16 ounces

Examples

Dimensional Analysis

1. How many seconds are there in 2.3 days?

$$2.3\ \text{days} \times \frac{24\ \text{hours}}{1\ \text{day}} \times \frac{60\ \text{minutes}}{1\ \text{hour}} \times \frac{60\ \text{seconds}}{1\ \text{minute}} = 198{,}720\ \text{seconds}$$

2. Calculate the number of milliliters in 10.2 gallons of gasoline.

$$10.2\ \text{gallons} \times \frac{4\ \text{quarts}}{1\ \text{gallon}} \times \frac{0.95\ \text{liter}}{1\ \text{quart}} \times \frac{1{,}000\ \text{ml}}{1\ \text{liter}} = 38{,}760\ \text{milliliters}$$

3. How many 500-mg extra-strength Tylenol tablets equal the maximum recommended dose of 1 gram?

$$1\ \text{gram} \times \frac{1{,}000\ \text{mg}}{1\ \text{gram}} \times \frac{1\ \text{tablet}}{500\ \text{mg}} = 2\ \text{tablets}$$

EXERCISE

Use dimensional analysis to solve these problems. Remember that numbers in the numerator should be preceded by the multiplication key, whereas numbers in the denominator should be preceded by the division key.

1. If you weigh 150 pounds, how many kilograms is that?
2. A tumor is reported to be 12 millimeters across. How many inches is that?
3. How many grams equal 1 ounce?
4. A football team has to advance the ball 10 yards. How many meters is that?
5. How many miles equal 1 kilometer?
6. At the deli, the roast beef you ordered weighs in at 1.85 pounds. How many ounces is that?

Answers

1. $$150 \text{ pounds} \times \frac{1 \text{ kilogram}}{2.2 \text{ pounds}} = 68.2 \text{ kilograms}$$

2. $$12 \text{ mm} \times \frac{1 \text{ cm}}{10 \text{ mm}} \times \frac{1 \text{ inch}}{2.54 \text{ cm}} = 0.47 \text{ inch}$$

3. $$1 \text{ ounce} \times \frac{1 \text{ pound}}{16 \text{ ounces}} \times \frac{454 \text{ grams}}{1 \text{ pound}} = 28.38 \text{ grams}$$

4. $$10 \text{ yards} \times \frac{3 \text{ feet}}{1 \text{ yard}} \times \frac{12 \text{ inches}}{1 \text{ foot}} \times \frac{1 \text{ meter}}{39.37 \text{ inches}} = 9.1 \text{ meters}$$

5. $$1 \text{ km} \times \frac{1{,}000 \text{ m}}{1 \text{ km}} \times \frac{39.37 \text{ inches}}{1 \text{ meter}} \times \frac{1 \text{ foot}}{12 \text{ in}} \times \frac{1 \text{ mile}}{5{,}280 \text{ ft}} = 0.62 \text{ mile}$$

6. $$1.85 \text{ pounds} \times \frac{16 \text{ ounces}}{1 \text{ pound}} = 29.6 \text{ ounces}$$

Because moles are a new idea, dimensional analysis will be useful to you in solving mole problems. You can rely on units and their cancelation in setting up problems correctly. Another advantage is that this approach works with any kind of problem involving units and their numbers.

Dimensional Analysis and Mole Problems

From what you have learned about moles to this point, you can write six mole conversion factors. Remember that you can write two factors for each (one the reciprocal of the other).

EXERCISES

Two of the six mole conversion factors are shown below. Write the other four.

1. $\dfrac{\text{1 mole}}{\text{molar mass in grams}}$
2.
3. $\dfrac{\text{1 mole}}{6.02 \times 10^{23} \text{ atoms or molecules}}$
4.
5.
6.

Answers

2. $\dfrac{\text{molar mass in grams}}{\text{1 mole}}$

4. $\dfrac{6.02 \times 10^{23} \text{ atoms or molecules}}{\text{1 mole}}$

5. $$\frac{\text{molar mass in grams}}{6.02 \times 10^{23}\text{ molecules or atoms*}}$$

6. $$\frac{6.02 \times 10^{23}\text{ molecules or atoms*}}{\text{molar mass in grams}}$$

*Use the word *atoms* when describing a single element; use the word *molecules* when describing molecules and compounds.

Strategy for Solving Mole Problems

- Begin with the *given*, which is the number provided in the problem. Be certain to include the accompanying unit of measure.
- From your list of six conversion factors, choose the one that has the same units in the denominator as does the given so that the two units will cancel.
- If the only units remaining after canceling are the units to be used in the answer, you are finished with the problem setup and need only to do the calculator work.
- If the remaining units are not the units in the answer, you will need another conversion factor. To choose this conversion factor, look for one with units in the denominator that are the same as the remaining units. Cancel all units that will cancel. If the remaining units are not in the answer, repeat this step.

Examples

Mole Problems

As you study these examples, examine how the order of the conversion factors matches the directions in the strategy plan shown above.

1. If there are 23 moles of water in a barrel, how many grams are there?

$$23\ \cancel{\text{moles of}}\ \text{water} \times \frac{18\text{ grams of }H_2O}{1\ \cancel{\text{mole}}\ H_2O} = 414\text{ grams }H_2O$$

2. How many molecules of water are there in 25 moles of water?

$$25\ \cancel{\text{moles}}\ \text{water} \times \frac{6.02 \times 10^{23}\text{ molecules of water}}{1\ \cancel{\text{mole}}\ \text{water}} = 1.505 \times 10^{25}\text{ molecules of water}$$

Note: If grams are not mentioned in the problem, molar mass should not be used.

3. If there are 5,000 molecules of CO_2, how many moles are there?

$$5{,}000 \text{ molecules } CO_2 \times \frac{1 \text{ mole } CO_2}{6.02 \times 10^{23} \text{ molecules}} = 8.3 \times 10^{-21} \text{ mole}$$

4. How many water molecules are in 1,000 grams of water?

$$1{,}000 \text{ grams } H_2O \times \frac{1 \text{ mole } H_2O}{18 \text{ grams}} \times \frac{6.02 \times 10^{23} \text{ molecules}}{1 \text{ mole } H_2O} = 3.34 \times 10^{25} \text{ molecules of water}$$

5. How many grams of NaOH (the active ingredient in Drano) are needed to have 3 moles?

$$3 \text{ moles NaOH} \times \frac{40 \text{ grams NaOH}}{1 \text{ mole NaOH}} = 120 \text{ grams NaOH}$$

EXERCISES

These problems cover all varieties of mole problems. As you set them up and enter them into your calculator, remember that molar mass must be calculated first only for problems in which grams are mentioned.

1. If 5,000,000 chlorine atoms are used to disinfect water, how many moles is that?
2. Thirty moles of carbon monoxide (CO) could be a lethal amount. How many grams is that?
3. If a nail contains 1.5 moles of iron, how many grams is that?
4. A roll of aluminum foil weighs 96 grams. How many moles of Al is that?
5. One liter of water weighs 1,000 grams. How many water molecules is that?
6. How many grams of caffeine ($C_8H_{10}N_4O_2$) are necessary to have 1,000,000 caffeine molecules?
7. The primary ingredient in Tums is calcium carbonate ($CaCO_3$). How many moles are in a 0.75-gram tablet?
8. If there is 4.2 moles of neon (Ne) gas in a restaurant sign, how many neon atoms are there?
9. Baking soda ($NaHCO_3$) is part of the recipe for biscuits. How many molecules of $NaHCO_3$ are there in 25 grams?
10. Nitrogen (N_2) is responsible for the bends, a potentially fatal problem associated with deep-sea diving. How many moles of N_2 are there in 9.5×10^{18} molecules of N_2?

11. Grain alcohol is C_2H_5OH (or C_2H_6O). How many grams does it take to equal 5.05 moles?

12. Ozone, (O_3) is considered an atmospheric "good guy" because of its ability to protect the earth from harmful ultraviolet radiation from the sun. What is the weight of 6.3×10^{12} molecules of ozone?

13. Carbonated beverages contain H_2CO_3. If there are 2.1 grams of it in a soda, how many moles are there?

14. Boric acid (H_3BO_3) is an ingredient in some eye wash preparations. How many molecules of H_3BO_3 are in 1.0 gram of boric acid?

15. The mercury ore cinnabar (HgS) was used to make the bright red paint used by Renaissance painters. What is the weight of 50 molecules of cinnabar?

Answers

1. $$5{,}000{,}000\ \cancel{\text{atoms}}\ \text{Cl} \times \frac{1\ \text{mole Cl}}{6.02\times10^{23}\ \cancel{\text{atoms}}} = 8.3\times10^{23}\ \text{mole Cl}$$

2. $$30\ \cancel{\text{moles}}\ \text{CO} \times \frac{28\ \text{grams CO}}{1\ \cancel{\text{mole}}\ \text{CO}} = 840\ \text{grams CO}$$

3. $$1.5\ \cancel{\text{moles}}\ \text{Fe} \times \frac{56\ \text{grams Fe}}{1\ \cancel{\text{mole Fe}}} = 84\ \text{grams Fe}$$

4. $$96\ \cancel{\text{grams}}\ \text{Al} \times \frac{1\ \text{mole Al}}{27\ \cancel{\text{grams Al}}} = 3.6\ \text{moles Al}$$

5. $$1{,}000\ \cancel{\text{grams}}\ H_2O \times \frac{1\ \cancel{\text{mole}}\ H_2O}{18\ \cancel{\text{grams}}} \times \frac{6.02\times10^{23}\ \text{molecules}}{1\ \cancel{\text{mole}}\ H_2O} = 3.3\ \times10^{23}\ \text{molecules}\ H_2O$$

6. $$1{,}000{,}000\ \cancel{\text{molecules}} \times \frac{1\ \text{mole}}{6.02\times10^{23}\ \cancel{\text{molecules}}} \times \frac{194\ \text{grams caffeine}}{1\ \text{mole}} = 3.22\times10^{-16}\ \text{gram of caffeine}$$

7. $$0.75\ \cancel{\text{gram}}\ CaCO_3 \times \frac{1\ \text{mole}\ CaCO_3}{100\ \cancel{\text{grams}}}\ 0.0075\ \text{mole}\ CaCO_3$$

8. $4.2 \text{ moles } N_2 \times \dfrac{6.02\times10^{23} \text{ molecules}}{1 \text{ mole}} = 2.53\times10^{24} \text{ molecules of } N_2$

9. $2.5 \text{ g } NaHCO_3 \times \dfrac{1 \text{ mole}}{84 \text{ grams}} \times \dfrac{6.02\times10^{23} \text{ molecules}}{1 \text{ mole}} = 1.79\times10^{23} \text{ molecules of } NaHCO_3$

10. $9.5 \times10^{18} \text{ molecules} \times \dfrac{1 \text{ mole}}{6.02\times10^{23} \text{ molecules}} = 1.58\times10^{-5} \text{ mole}$

11. $5.05 \text{ moles} \times \dfrac{46 \text{ grams } C_2H_5OH}{1 \text{ mole}} = 232.30 \text{ grams } C_2H_5OH$

12. $6.3 \times10^{12} \text{ molecules} \times \dfrac{1 \text{ mole}}{6.02\times10^{23} \text{ molecules}} \times \dfrac{48 \text{ grams } O_3}{1 \text{ mole}} = 5.03\times10^{-10} \text{ gram of } O_3$

13. $2.1 \text{ grams } H_2CO_3 \times \dfrac{1 \text{ mole}}{62 \text{ grams}} = 0.03 \text{ mole } H_2CO_3$

14. $1 \text{ gram } H_3BO_3 \times \dfrac{1 \text{ mole}}{62 \text{ grams}} \times \dfrac{6.02\times10^{23} \text{ molecules}}{1 \text{ mole}} = 9.71\times10^{21} \text{ molecules } H_3BO_3$

15. $50 \text{ molecules } HgS \times \dfrac{1 \text{ mole}}{6.02\times10^{23} \text{ molecules}} \times \dfrac{233 \text{ grams}}{1 \text{ mole}} = 1.94\times10^{-20} \text{ gram of } HgS$

UNIT 5

Stoichiometry

Stoichiometry, Chemical Recipes, an Integration of Moles, and Balanced Equations

Stoichiometry is the technical word for the relationships among balanced equations, moles, and grams. Stoichiometry is to chemists what cooking and recipes are to cooks.

The balanced equation that we will use for this stoichiometry explanation is the recipe for the manufacture of ammonia (NH_3). This reaction was so important that the chemist responsible for it, Fritz Haber, was awarded the Nobel Prize. Ammonia is a gateway step in the manufacture of fertilizers, and its manufacture was a giant step in solving the problem of providing food to a world population growing at an exponential rate. The equation for the Haber process is

$$3H_2 \quad + \quad N_2 \quad \rightarrow \quad 2NH_3$$

The coefficients in all balanced equations are about counting and therefore are about moles and other counting numbers. In the above equation you could say:

3 molecules of H_2	plus	1 molecule of N_2	yields	2 molecules of NH_3
3 dozen H_2	plus	1 dozen N_2	yields	2 dozen NH_3
3 moles H_2	plus	1 mole N_2	yields	2 moles NH_3

Notice that you cannot use the unit *gram* here. The coefficients do not refer to weighing but rather to counting. Once again, the mole–molar mass relationship is involved.

Balanced equation coefficients provide you with another set of conversion factors.

For the specific equation above, you can say:

$$\frac{3 \text{ moles } H_2}{1 \text{ mole } N_2} \quad \text{and} \quad \frac{1 \text{ mole } N_2}{3 \text{ moles } H_2}$$

EXERCISE

Four more conversion factors can be written from the balanced equation coefficients for this equation. What are they?

1.
2.
3.
4.

Answers

$$\frac{2 \text{ moles } NH_3}{3 \text{ moles } H_3} \qquad \frac{3 \text{ moles } H_2}{2 \text{ moles } NH_3} \qquad \frac{2 \text{ moles } NH_3}{1 \text{ mole } N_2} \qquad \frac{1 \text{ mole } N_2}{2 \text{ moles } NH_3}$$

These mole relationships are used to move from one reactant or product of the balanced equation to another.

STRATEGY FOR SOLVING STOICHIOMETRY PROBLEMS

- Begin by writing the given, using the number as well as the units and including the chemical formula. *Hint:* The given is the number and the unit of measure provided in the problem.
- The next conversion factor should have in its denominator the same units as in the given.
- If the numerator of this conversion factor has the units needed in the solution, then you are finished with the setup and need only to do the arithmetic involved.
- If you still do not have the units needed in the answer, you will require at least one more conversion factor. No problem requires more than three conversion factors.

- Although there are four kinds of stoichiometry problems, the game plan shown here works for all of them.

Mole-to-Mole Problems

If you want to manufacture 150 moles of ammonia, how many moles of hydrogen will you need? (Remember to follow the dimensional analysis problem-solving strategy in the previous chapter).

$$150 \text{ moles } NH_3 \times \frac{3 \text{ moles } H_2}{2 \text{ moles } NH_3} = 225 \text{ moles } H_2$$

As you look at this solution, the points to notice are

- Begin with the given number provided in the problem, including its units.
- Choose a conversion factor that has in its denominator the same units as those accompanying the given number.
- If possible, the numerator of this conversion factor should contain the units that you are looking for in the answer.
- Cancel out all the units that you can.
- The units that remain must be the same as the units that you are looking for the answer.

EXERCISES

Use this equation to solve the following mole-to-mole problems.

$$3H_2 \quad + \quad N_2 \quad \rightarrow \quad 2NH_3$$

1. If you want to make 100 moles of ammonia (NH_3), how many moles of nitrogen (N_2) will you need?
2. If you have 12 moles of hydrogen, how many moles of ammonia can you make?
3. If you have 36 moles of nitrogen, how many moles of hydrogen will you need? How many moles of ammonia will that make?

Answers

1. $$100 \text{ moles } NH_3 \times \frac{1 \text{ mole } N_2}{2 \text{ moles } NH_3} = 50 \text{ moles } N_2$$

2. $$12 \text{ moles } H_2 \times \frac{2 \text{ moles } NH_3}{3 \text{ moles } H_2} = 8 \text{ moles } NH_3$$

3. $$36 \text{ moles } N_2 \times \frac{3 \text{ moles } H_2}{1 \text{ mole } N_2} = 108 \text{ moles } H_2$$

$$108 \text{ moles } H_2 \times \frac{2 \text{ moles } NH_3}{3 \text{ moles } H_2} = 72 \text{ moles } NH_3$$

In the real world, grams are discussed more frequently than moles. As the chief executive officer of a chemical company you would not see orders for a certain number of moles of a material, but rather for grams, tons, or some other weight unit.

Gram-to-Gram Problems

The balanced equation below describes the process for etching glass (SiO_2) by treating it with the highly corrosive hydrofluoric acid (HF).

$$4HF \quad + \quad SiO_2 \quad \rightarrow \quad SiF4 \quad + \quad 2H_2O$$

How many grams of glass (SiO_2) can be etched by 100 grams of hydrofluoric acid (HF)? Following the strategy above,

$$100 \text{ grams HF} \times \frac{1 \text{ mole HF}}{20 \text{ grams HF}} \times \frac{1 \text{ mole } SiO_2}{4 \text{ moles HF}} \times \frac{60 \text{ grams } SiO_2}{1 \text{ mole } SiO_2} = 75 \text{ grams } SiO_2$$

Note that

- The solution begins with the only number given in the problem, along with the units and the formula (HF) associated with that number.
- The next conversion factor must have grams of HF in the denominator. It is the molar mass of HF.
- The third conversion factor has moles given in its denominator and the moles asked for in its numerator. These numbers are the coefficients from the balanced equation.

- The fourth conversion factor allows for the transition from moles of SiO_2 to grams of SiO_2 via the molar mass.
- This four-step solution is always used for any problem in which grams are given and grams are asked for.

EXERCISES

Use the strategy outlined above to solve these gram-to-gram problems.

1. Miners and spelunkers use a lamp fueled by water combined with calcium carbide (CaC_2) to produce the flammable acetylene (C_2H_2) and the waste product calcium hydroxide ($Ca(OH)_2$):

$$CaC_2 \quad + \quad 2H_2O \quad \rightarrow \quad Ca(OH)_2 \quad + \quad C_2H_2$$

How many grams of water are required to react with 100 grams of calcium carbide?

2. How many grams of oxygen (O_2) are required for the burning of 500 grams of propane (C_3H_8) according to the following balanced equation?

$$C_3H_8 \quad + \quad 5O_2 \quad \rightarrow \quad 3CO_2 \quad + \quad 4H_2O$$

3. For the balanced equation in question 2, how many grams of water are made at the same time that 250 grams of carbon dioxide (CO_2) are produced?

Answers

1. $$100 \text{ grams } CaC_2 \times \frac{1 \text{ mole } CaC_2}{64 \text{ grams } CaC_2} \times \frac{2 \text{ moles } H_2O}{1 \text{ mole } CaC_2} \times \frac{18 \text{ grams } H_2O}{1 \text{ mole } H_2O} = 56.25 \text{ grams } H_2O$$

2. $$500 \text{ grams } C_3H_8 \times \frac{1 \text{ mole } C_3H_8}{44 \text{ grams}} \times \frac{5 \text{ moles } O_2}{1 \text{ mole } C_3H_8} \times \frac{32 \text{ grams } O_2}{1 \text{ mole } O_2} = 1{,}818.2 \text{ grams } O_2$$

3. $$250 \text{ grams } CO_2 \times \frac{1 \text{ mole } CO_2}{44 \text{ grams}} \times \frac{4 \text{ moles } H_2O}{3 \text{ moles } CO_2} \times \frac{18 \text{ grams } H_2O}{1 \text{ mole } H_2O} = 136.4 \text{ grams } H_2O$$

The kind of stoichiometry problems that you have just worked through are called gram-to-gram problems, with grams in both the given and in the solution. Solving gram-to-gram problems always involves the same four-step strategy illustrated above. All other kinds of stoichiometry problems are shorter by one or two conversion factors.

At the end of this unit there is a comprehensive problem set that will give you the opportunity for more practice.

Mole-to-Gram Problems

Look at the four-step problem-solving format above. A Mole-to-gram problem is solved in the same way, except that you write the given followed by the last two conversion factors. Therefore, it is a three-step problem rather than a four-step problem.

As an example, we use the previous balanced equation:

$$C_3H_8 \quad + \quad 5O_2 \quad \rightarrow \quad 3CO_2 \quad + \quad 4H_2O$$

How many grams of water are produced at the same time that 100 moles of CO_2 are made.

$$100 \text{ moles } CO_2 \times \frac{4 \text{ moles } H_2O}{3 \text{ moles } CO_2} \times \frac{18 \text{ grams } H_2O}{1 \text{ mole } H_2O} = 2{,}400 \text{ grams } H_2O$$

EXERCISES

For all these mole-to-gram problems, use the following balanced equation and the strategy discussed above.

$$2Al_2O_3 \quad + \quad 3C \quad \rightarrow \quad 4Al \quad + \quad 3CO_2$$

1. How many grams of aluminum (Al) can be produced from 75 moles of aluminum oxide (Al_2O_3)?
2. If there are 9 moles of carbon (C), how many grams of Al_2O_3 are needed?
3. At the same time that 300 moles of carbon dioxide (CO_2) are produced, how many grams of aluminum are also produced?

Answers

1. $$75\ \text{moles}\ Al_2O_3 \times \frac{4\ \text{moles Al}}{2\ \text{moles}\ Al_2O_3} \times \frac{27\ \text{grams Al}}{1\ \text{mole Al}} = 4{,}050\ \text{grams Al}$$

2. $$9\ \text{moles C} \times \frac{2\ \text{moles}\ Al_2O_3}{3\ \text{moles C}} \times \frac{102\ \text{grams}\ Al_2O_3}{1\ \text{mole}\ Al_2O_3} = 612\ \text{grams}\ Al_2O_3$$

3. $$300\ \text{moles}\ CO_2 \times \frac{4\ \text{moles Al}}{3\ \text{moles}\ CO_2} \times \frac{27\ \text{grams Al}}{1\ \text{mole Al}} = 10{,}800\ \text{grams Al}$$

Up to this point, we have dealt with mole-to-mole, gram-to-gram and mole-to-gram problems. That leaves us with gram-to-mole problems.

Gram-to-Mole Problems

As an example of gram-to-mole problem solving, we use the following equation:

$$4NH_3 \quad + \quad 5O_2 \quad \rightarrow \quad 4NO \quad + \quad 6H_2$$

As you study this example, look back at the four-step strategy for solving gram-to-gram problems. Note that we will follow the same pattern except for omitting the last conversion factor.

How many moles of NO are produced at the same time as 100 grams of H_2?

$$100\ \text{grams}\ H_2 \times \frac{1\ \text{mole}\ H_2}{2\ \text{grams}} \times \frac{4\ \text{moles NO}}{6\ \text{moles}\ H_2} = 33.3\ \text{moles NO}$$

EXERCISES

Solve the following gram-to-mole problems.

1. How many moles of ammonia (NH_3) are needed to react with 5 grams of oxygen (O_2)?
2. In order to get 400 grams of NO, how many moles of oxygen (O_2) are needed?
3. How many moles of H_2 would be the result of reacting 200 grams of O_2?

Answers

1. $$5 \text{ grams } O_2 \times \frac{1 \text{ mole } O_2}{32 \text{ grams}} \times \frac{4 \text{ moles } NH_3}{5 \text{ moles } O_2} = 0.13 \text{ mole } NH_3$$

2. $$400 \text{ grams NO} \times \frac{1 \text{ mole NO}}{30 \text{ grams}} \times \frac{5 \text{ moles } O_2}{4 \text{ moles NO}} = 16.57 \text{ moles } O_2$$

3. $$200 \text{ grams } O_2 \times \frac{1 \text{ mole } O_2}{32 \text{ grams}} \times \frac{6 \text{ moles } H_2}{5 \text{ moles } O_2} = 7.5 \text{ moles } H_2$$

LIMITING FACTORS

Limiting factors can be explained by thinking through what is involved in sandwich making. By considering the sandwich ingredients you have on hand, you can calculate the number of sandwiches you will be able to make. One of the sandwich ingredients limits the number of sandwiches you can make and therefore is the limiting factor.

For example, suppose that you have 10 slices of bologna and 18 slices of bread. The bread supply allows you to make 9 sandwiches with 1 slice of bologna left over. The bread is the *limiting factor,* and the bologna is the *excess.* All limiting-factor problems are like this example in that you are given numerical values for each reactant and are asked to calculate the amount of product (in this case, sandwiches) that can be made. Sometimes you are asked to calculate the amount of excess reactant (ingredient) that is unused. In this case, the 1 slice of bologna is the excess reactant that is unused.

In a limiting-factor problem two amounts are given. It is necessary to decide which of the two will control, or limit, the amount of product that can be made.

A Limiting-Factor Problem

The idea of a limiting factor will be explored by using an equation involving automobiles. You have likely heard that jump-starting a car battery can be dangerous because hydrogen and oxygen can be produced and form a potentially explosive combination. The balanced equation for this reaction is

$$2H_2 \quad + \quad O_2 \quad \rightarrow \quad 2H_2O$$

Suppose that you have 3.0 grams of hydrogen (H_2) and 32 grams of oxygen (O_2). Which is the limiting factor? Remember that the coefficients in a balanced equation do not refer to grams but rather to moles. The first step in a limiting-factor problem is to change all grams to moles.

$$3.0 \text{ grams } H_2 \times \frac{1 \text{ mole } H_2}{2 \text{ grams } H_2} = 1.5 \text{ moles } H_2$$

$$32 \text{ grams } O_2 \times \frac{1 \text{ mole } O_2}{32 \text{ grams } O_2} = 1.0 \text{ mole } O_2$$

Now compare the number of moles that you have with the number of moles in the balanced equation. If you use the 1.5 moles of H_2, you will need 0.75 mole of O_2 because of the 2:1 ratio in the balanced equation. This plan works, with 0.25 mole of O_2 left over. The O_2 is the excess, and the H_2 is the limiting factor.

This leads to another truth: Whichever reactant works is the limiting factor.

Once you know that the hydrogen is the limiting factor, there is no need to examine the oxygen. However, just to make certain, it is a good idea to look at the oxygen. If you use 1.0 mole of oxygen, you will need 2.0 moles of hydrogen. This is the stumbling block because you have only 1.5 moles of hydrogen. This means that using all the oxygen does not work, and so the hydrogen is the limiting factor.

You must arrive at the same conclusion as to the identity of the limiting factor no matter which approach you use. Typical limiting-factor stoichiometry problems ask for three things:

1. Which reactant is the limiting factor?
2. How much product is produced?
3. How much excess is left over?

In this example:

- Hydrogen is the limiting factor.
- The amount of product is calculated by the same method used earlier for mole-to-gram stoichiometry problems. Start with the moles of the limiting factor because a limiting factor is defined as the reactant that limits or determines the amount of product that can be made.

$$1.5 \text{ moles } H_2O \times \frac{18 \text{ grams } H_2O}{1 \text{ mole } H_2O} = 27.0 \text{ grams } H_2O$$

The amount of excess left over is determined by subtracting the amount used from the amount given. In this case, oxygen is in excess (because hydrogen is the limiting factor). The given amount of oxygen is 1.0 mole. The amount used is 0.75 mole, and so the amount left over is 1.0 – 0.75 = 0.25 mole.

EXERCISE

For each of these problems find out which reactant is the limiting factor. Then calculate the number of grams of water made and the number of moles of excess. Use the following balanced equation for the combustion involved in a propane heater.

$$C_3H_8 \quad + \quad 5O_2 \quad \rightarrow \quad 3CO_2 \quad + \quad 4H_2O$$

1. For 100 moles of each reactant, which is the limiting factor, how many grams of water are produced, and how many grams of excess remain?
2. Perform the same three calculations when there are 100 grams of each reactant.
3. Perform the same three calculations when there are 110 grams of propane (C_3H_8) and 480 grams of oxygen (O_2).

Answers

1. If all 100 moles of propane are used, 500 moles of oxygen are needed. With only 100 moles of oxygen available, this would not work. If all 100 moles of oxygen are used, the amount of propane that would react with it is

$$100 \ \cancel{\text{moles } O_2} \times \frac{1 \text{ mole } C_3H_8}{5 \ \cancel{\text{moles } O_2}} = 20 \text{ moles } C_3H_8$$

With 100 moles of C_3H_8 available, this would work and so oxygen is the limiting factor.

Once you know that O_2 is the limiting factor, use its number of moles to calculate the amount of product.

$$100 \ \cancel{\text{moles } O_2} \times \frac{4 \text{ moles } H_2O}{5 \ \cancel{\text{moles } O_2}} \times \frac{18 \text{ grams } H_2O}{1 \ \cancel{\text{mole } H_2O}}$$

Find the amount of excess:

100 moles of C_3H_8 given – 20 moles of C_3H_9 used
(see limiting-factor calculation) = 80 moles of C_3H_8 left over

2. Find the limiting factor. Change the grams of the given reactants to moles:

$$100 \text{ grams of } C_3H_8 \times \frac{1 \text{ mole } C_3H_8}{44 \text{ g } C_3H_8} = 2.3 \text{ moles } C_3H_8$$

$$100 \text{ grams of } O_2 \times \frac{1 \text{ mole } O_2}{32 \text{ g } O_2} = 3.1 \text{ moles } O_2$$

In the balanced equation, for each mole of C_3H_8, 5 moles of O_2 are needed. If all the C_3H_8 is used, 5×2.3 moles of O_2 are required. Because that much O_2 is not available, the C_3H_8 cannot all be used, and therefore it is not the limiting factor. The O_2 must be the limiting factor.

Find the amount of water that is made, begin with the moles of the limiting factor:

$$3.1 \text{ moles } O_2 \times \frac{4 \text{ moles } H_2O}{5 \text{ moles } O_2} \times \frac{18 \text{ grams } H_2O}{1 \text{ mole } H_2O} = 44.6 \text{ grams } H_2O$$

Find the amount of the excess left over:

2.3 moles C_3H_8 given
–0.6 mole C_3H_8 used (one-fifth of that given
because of the 1:5 coefficient ratio)
= 1.7 moles C_3H_8 leftover

3. $$110 \text{ grams } C_3H_8 \times \frac{1 \text{ mole } C_3H_8}{44 \text{ g } C_3H_8} = 2.5 \text{ moles } C_3H_8$$

$$480 \text{ grams } O_2 \times \frac{1 \text{ mole } O_2}{32 \text{ grams } O_2} = 15 \text{ moles } O_2$$

Check to see if all the propane can be used.

$$2.5 \text{ moles } C_3H_8 \times \frac{5 \text{ moles } O_2}{1 \text{ mole } C_3H_8} = 12.5 \text{ moles } O_2$$

This means that all 2.5 moles of propane can be used, because there are 15 moles of oxygen. Because this works, propane is the limiting factor.

Calculate the grams of water that can be produced:

$$2.5 \text{ moles } C_3H_8 \times \frac{4 \text{ moles } H_2O}{1 \text{ mole } C_3H_8} \times \frac{18 \text{ grams } H_2O}{1 \text{ mole } H_2O} = 180 \text{ grams } H_2O$$

Calculate the leftover O_2:

$$15.0 \text{ moles } O_2 \text{ given} - 12.5 \text{ moles } O_2 \text{ used} = 2.5 \text{ moles } O_2 \text{ leftover}$$

GENERAL REVIEW OF STOICHIOMETRY

The following set of exercises includes each type of stoichiometry problem. They are presented in random order and not grouped by type.

EXERCISES

Be certain to balance the equation before solving each of these problems.

1. $HNO_3 + NaOH \rightarrow NaNO_3 + H_2O$

   How many moles of nitric acid (HNO_3) are required to make 50 grams of sodium nitrate ($NaNO_3$)?

2. $Zn + HCl \rightarrow ZnCl_2 + H_2$

   How many moles of zinc are required to react with 2.5 moles of hydrochloric acid (HCl)?

3. $Al_2O_3 \rightarrow Al + O_2$

   How many grams of aluminum (Al) can be made from 500 grams of Al_2O_3?

4. $SO_2 + O_2 \rightarrow SO_3$

   How many grams of O_2 are required to react with 0.75 mole of sulfur dioxide (SO_2)? (*Hint:* When calculating the molar mass of O_2, remember to use the oxygen twice.)

5. $H_2O \rightarrow H_2 + O_2$

   How many grams of oxygen can be produced from 50 grams of water?

6. $H_2SO_4 + Ca(OH)_2 \rightarrow H_2O + CaSO_4$

   How many grams of calcium sulfate can be made from 100 grams of calcium hydroxide [$Ca(OH)_2$]?

7. $N_2 + H_2 \rightarrow NH_3$

   Which reactant is the limiting factor if there are 56 grams of nitrogen and 14 grams of hydrogen? How many grams of ammonia (NH_3) can be produced?

8. $C_4H_{10} + O_2 \rightarrow CO_2 + H_2O$

How many moles of oxygen are required to react with 7 moles of butane (C_4H_{10})?

9. $Co + O_2 \rightarrow Co_2O_3$

Beginning with 7.2 moles of cobalt (Co) would result in how many grams of cobalt(III) oxide (Co_2O_3)?

10. $Al + HNO_3 \rightarrow H_2 + Al(NO_3)_3$

How many grams of nitric acid (HNO_3) are needed to react with 80 grams of aluminum (Al)?

11. $PBr_3 + H_2O \rightarrow HBr + H_3PO_3$

How many grams of hydrobromic acid (HBr) can be made from 17 moles of phosphorus tribromide (PBr_3)?

12. $NH_3 + Cl_2 \rightarrow NCl_3 + NH_4Cl$

How many moles of ammonia (NH_3) are needed to make 18 moles of ammonium chloride (NH_4Cl)?

13. $NO + O_2 \rightarrow NO_2$

If you began with 100 grams of both reactants, which will be the limiting factor?

Then how many grams of NO_2 can be made?

14. $AgNO_3 + Zn \rightarrow Zn(NO_3)_2 + Ag$

How many grams of silver (Ag) can be made from 100 grams of zinc (Zn)?

15. $Mg + O_2 \rightarrow MgO$

How many moles of O_2 are needed to react with 60 grams of magnesium (Mg)?

16. $N_2 + H_2 \rightarrow NH_3$

How many moles of hydrogen are required to make 19.7 moles of NH_3?

17. $Ba(OH)_2 \rightarrow BaO + H_2O$

In order to make 200 grams of BaO, how many grams of barium hydroxide [$Ba(OH)_2$] should you begin with?

18. $CH_4 + O_2 \rightarrow CO_2 + H_2O$

How many grams of water are made at the same time that 17 moles of carbon dioxide are produced?

19. $PbO + HCl \rightarrow PbCl_2 + H_2O$

How many moles of hydrochloric acid (HCl), are needed to react with 500 grams of PbO?

20. $P_4 + H_2 \rightarrow PH_3$

How many grams of PH_3 can be made from 0.24 gram of hydrogen and 4 grams of phosphorus?

Answers

1. $$50 \text{ grams } NaNO_3 \times \frac{1 \text{ mole } NaNO_3}{85 \text{ grams } NaNO_3} \times \frac{1 \text{ mole } HNO_3}{1 \text{ mole } NaNO_3} = 0.59 \text{ mole } HNO_3$$

2. $$2.5 \text{ moles HCl} \times \frac{1 \text{ mole Zn}}{2 \text{ moles HCl}} = 1.25 \text{ moles Zn}$$

3. $$500 \text{ grams } Al_2O_3 \times \frac{1 \text{ mole } Al_2O_3}{102 \text{ grams } Al_2O_3} \times \frac{4 \text{ moles Al}}{2 \text{ moles } Al_2O_3} \times \frac{27 \text{ grams Al}}{1 \text{ mole Al}} = 264.71 \text{ grams Al}$$

4. $$0.75 \text{ mole } SO_2 \times \frac{1 \text{ mole } O_2}{2 \text{ moles } SO_2} \times \frac{32 \text{ g } O_2}{1 \text{ mole } O_2} = 12 \text{ g } O_2$$

5. $$50 \text{ g } H_2O \times \frac{1 \text{ mole } H_2O}{18 \text{ g } H_2O} \times \frac{1 \text{ mole } O_2}{2 \text{ moles } H_2O} \times \frac{32 \text{ g } O_2}{1 \text{ mole } O_2} = 44.44 \text{ g } O_2$$

6. $$100 \text{ g } Ca(OH)_2 \times \frac{1 \text{ mole } Ca(OH)_2}{74 \text{ g } Ca(OH)_2} \times \frac{1 \text{ mole } CaSO_4}{1 \text{ mole } Ca(OH)_2} \times \frac{136 \text{ g } CaSO_4}{1 \text{ mole } CaSO_4} = 183.78 \text{ g } CaSO_4$$

7. $$56 \cancel{\text{g } N_2} \times \frac{1 \text{ mole } N_2}{28 \cancel{\text{g}} \ N_2} = 2 \text{ moles } N_2$$

$$14 \cancel{\text{g } H_2} \times \frac{1 \text{ mole } H_2}{2 \cancel{\text{g}} \ H_2} = 7 \text{ moles } H_2$$

To use all the N_2:

$$2 \cancel{\text{moles } N_2} \times \frac{3 \text{ moles } H_2}{1 \cancel{\text{mole } N_2}} = 6 \text{ moles } H_2 \text{ required}$$

This works. There will be 1 mole of H_2 left over if all the nitrogen is used. The hydrogen is the excess, and the nitrogen is the limiting factor.

To find the grams of NH_3 that can be made, start with the nitrogen, as it limits the amount of product that can be made.

$$2 \text{ moles } N_2 \times \frac{2 \text{ moles } NH_3}{1 \text{ mole } N_2} \times \frac{17 \text{ g } NH_3}{1 \text{ mole}} = 68 \text{ g } NH_3$$

8. $$7 \text{ moles } C_4H_{10} \times \frac{13 \text{ moles } O_2}{2 \text{ moles } C_4H_{10}} = 45.5 \text{ moles } O_2$$

9. $$7.2 \text{ moles Co} \times \frac{2 \text{ moles } Co_2O_3}{4 \text{ moles Co}} \times \frac{166 \text{ g } CoCO_3}{1 \text{ mole } CoCO_3} = 597.60 \text{ g } CoCO_3$$

10. $$80 \text{ g Al} \times \frac{1 \text{ mole Al}}{27 \text{ g Al}} \times \frac{2 \text{ moles } HNO_3}{1 \text{ mole Al}} \times \frac{63 \text{ g } HNO_3}{1 \text{ mole } HNO_3} = 373.33 \text{ g } HNO_3$$

11. $$17 \text{ moles PBr} \times \frac{3 \text{ moles HBr}}{1 \text{ mole } PBr_3} \times \frac{81 \text{ g HBr}}{1 \text{ mole HBr}} = 4{,}131 \text{ g HBr}$$

12. $$18 \text{ moles } NH_3 \times \frac{3 \text{ moles } NH_4Cl}{4 \text{ moles } NH_3} = 13.5 \text{ moles } NH_4Cl$$

13. This is a limiting-factor problem; so the first thing to do is to find the number of moles of both reactants:

$$100 \text{ g NO} \times \frac{1 \text{ mole NO}}{30 \text{ g NO}} = 3.30 \text{ moles NO}$$

$$100 \text{ g } O_2 \times \frac{1 \text{ mole } O_2}{32 \text{ g } O_2} = 3.13 \text{ moles } O_2$$

If all the NO is used, we will need half as many moles of O_2 because of the two-to-one balanced equation ratio. Half of the 3.30, or 1.65 moles, of O_2 is required. There is more than enough O_2, with 3.13 moles, so it would work to use all the NO. Then 3.13 – 1.65 = 1.48 moles of O_2 are left over.

Knowing that NO is the limiting factor, we calculate the grams of NO_2 that can be made:

$$3.3 \ \cancel{\text{moles NO}} \times \frac{2 \ \cancel{\text{moles NO}_2}}{2 \ \cancel{\text{moles NO}}} \times \frac{46 \text{ g NO}_2}{1 \ \cancel{\text{mole NO}_2}} = 151.3 \text{ g NO}_2$$

14. $$100 \ \cancel{\text{g Zn}} \times \frac{1 \ \cancel{\text{mole Zn}}}{65 \ \cancel{\text{g Zn}}} \times \frac{2 \ \cancel{\text{moles Ag}}}{1 \ \cancel{\text{mole Zn}}} \times \frac{108 \text{ g Ag}}{1 \ \cancel{\text{mole Ag}}} = 332.31 \text{ g Ag}$$

15. $$60 \ \cancel{\text{g Mg}} \times \frac{1 \ \cancel{\text{mole Mg}}}{24 \ \cancel{\text{g Mg}}} \times \frac{1 \text{ mole O}_2}{2 \ \cancel{\text{moles Mg}}} = 1.25 \text{ moles O}_2$$

16. $$19.7 \ \cancel{\text{moles NH}_3} \times \frac{3 \text{ moles H}_2}{2 \ \cancel{\text{moles NH}_3}} = 29.55 \text{ moles H}_2$$

17. $$200 \ \cancel{\text{g BaO}} \times \frac{1 \ \cancel{\text{mole BaO}}}{153 \ \cancel{\text{g BaO}}} \times \frac{1 \ \cancel{\text{mole Ba(OH)}_2}}{1 \ \cancel{\text{mole BaO}}} \times \frac{171 \text{ g Ba(OH)}_2}{1 \ \cancel{\text{mole}}} = 223.53 \text{ g Ba(OH)}_2$$

18. $$17 \ \cancel{\text{moles CO}_2} \times \frac{2 \ \cancel{\text{moles H}_2\text{O}}}{1 \ \cancel{\text{mole CO}_2}} \times \frac{18 \ \cancel{\text{g H}_2\text{O}}}{1 \ \cancel{\text{mole H}_2\text{O}}} = 612 \text{ g H}_2\text{O}$$

19. $$500 \ \cancel{\text{g PbO}} \times \frac{1 \ \cancel{\text{mole PbO}}}{223 \ \cancel{\text{g PbO}}} \times \frac{2 \text{ moles HCl}}{1 \ \cancel{\text{mole PbO}}} = 4.48 \text{ moles HCl}$$

20. There is a limiting-factor problem. Find the moles of each reactant first.

$$4 \text{ g P} \times \frac{1 \text{ mole P}}{31 \text{ g P}} = 0.03 \text{ mole P}$$

$$0.24 \text{ g } H_2 \times \frac{1 \text{ mole } H_2}{2 \text{ g } H_2} = 0.12 \text{ mole } H_2$$

If you were use all the phosphorus, you will need 6 times that amount of moles of hydrogen because of the 6:1 mole ratio in the balanced equation. Multiply all the phosphorus (0.03) by 6 = 0.18 mole of hydrogen needed. But you have only 0.12 mole of hydrogen, so you cannot use all the phosphorus. This means that hydrogen is the only one you can use totally, and it therefore is the limiting factor. Just to be certain, if you use all the 0.12 mole of hydrogen, you will need one-sixth of that amount in phosphorus. One-sixth of 0.12 is 0.02 mole of phosphorus. Because the moles of phosphorus present is 0.03, you will have 0.01 mole left over.

Because hydrogen is the limiting factor, begin with its amount.

$$0.12 \text{ mole } H_2 \times \frac{4 \text{ moles } PH_3}{6 \text{ moles } H_2} \times \frac{34 \text{ g } PH_3}{1 \text{ mole } PH_3} = 2.72 \text{ g } PH_3$$

Gases

Gas Theory, Problem Solving, Applications, Absolute Zero, and the Ideal Gas Law

Gases provide the breath of life, inflate auto tires, power hot-air balloons, dissolve in our blood, heat our homes, and even protect us with automobile air bags. The gases in our environment span the spectrum from necessity to usefulness to extreme toxicity. Carbon monoxide from faulty combustion in kerosene heaters is implicated in numerous deaths each year. The nitrous oxide that anesthetizes us in the dental chair is the same gas that is so dangerous when sniffed by drug addicts. The horrors of gas warfare stalk our terrorism preparation. The list of gases goes on and on, with the study of gases made interesting by their very widespread nature and practicality.

As you teach yourself about gases in this unit, develop a mental picture of how the molecules of a gas behave. The model for the action of gas molecules in a confined space is called *kinetic molecular theory*. Because *kinetic* means moving, the theory describes how molecules behave as a result of their motion. The speed of molecules is related to their temperature, in that the hotter they are, the faster they move. In an enclosed space, molecules randomly strike each other and collide with the walls of the container as they travel (Figure 6.1).

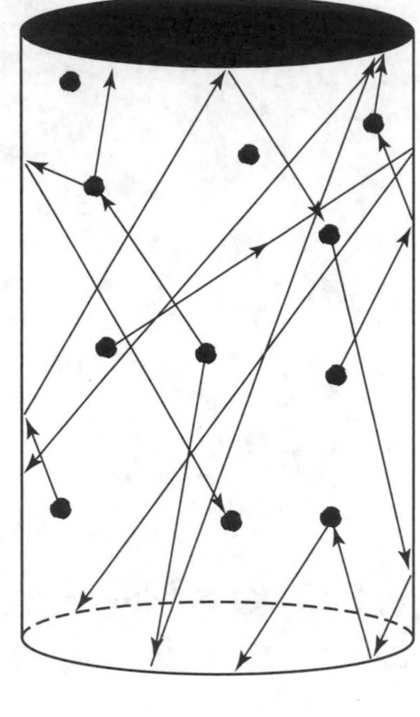

Figure 6.1: Kinetic molecular theory

PRESSURE

As the molecules of a gas in a container collide with the inside surfaces of the container, their speed and mass create a push against the walls. This push is called *pressure*. The helium atoms inside a helium-filled balloon exert enough pressure to keep the balloon inflated. The pressure inside an oxygen tank is the result of oxygen molecules striking the interior of the tank. In addition to the pressure of gases enclosed in a container, there is the pressure of the air around us.

Atmospheric Pressure

We live in a sea of air. This air has weight and exerts pressure on everything. We do not notice this pressure because it is the same inside our homes as outside them and the same inside our bodies as outside them. We do notice when atmospheric pressure changes quickly, as in an airplane or when traveling to a different altitude. The familiar ear-popping sensation makes us aware of these changes in atmospheric pressure.

Atmospheric pressure is also called *barometric pressure* and is measured by an instrument called a barometer, as shown in Figure 6.2. The weight of the atmosphere pushes down on the mercury, holding it up in the tube. When the atmospheric pressure increases, as it does with clear weather, the mercury rises in the tube. When the atmospheric pressure decreases, as it does with stormy weather, the mercury falls to a lower level in the tube. A factor in the destructive power of a hurricane is the extreme low pressure associated with it.

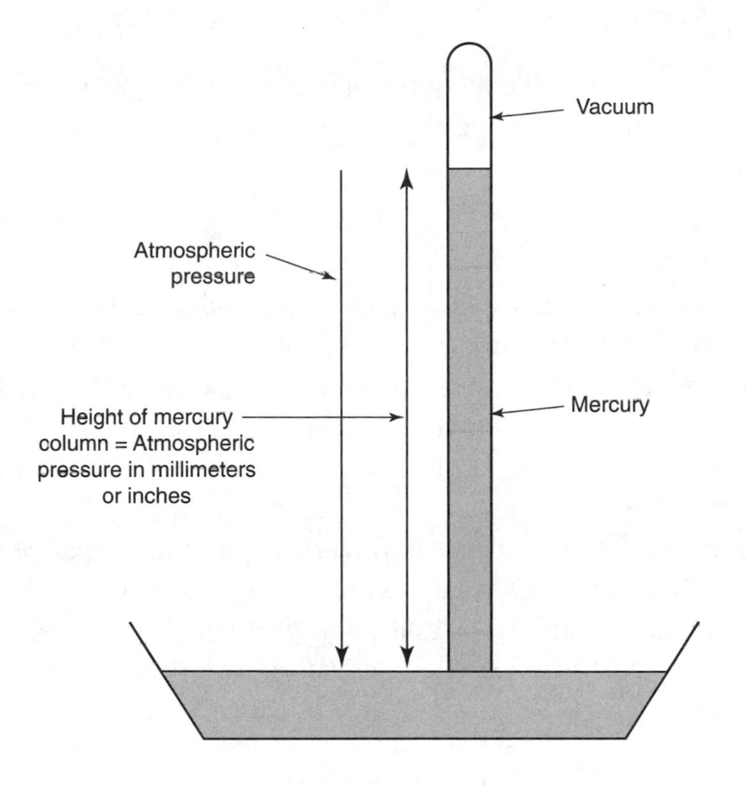

Figure 6.2: Barometer

Elevation is another factor influencing atmospheric pressure. As the elevation above sea level increases, the atmosphere "thins out" and exerts less pressure. Athletes sometimes train at higher elevations in order to accustom their bodies to functioning with less oxygen. They are then in better shape to perform in the oxygen-rich environments of lower elevations. Even the design of potato chip bags takes atmospheric pressure into account. A potato chip bag taken to a high elevation expands and becomes very tight. This is caused by the decreased pressure on the outside of the bag, resulting in the pressure inside pushing the bag outward.

Units of Pressure

A variety of pressure units are in use. These units are related to the column of mercury shown in Figure 6.2. The height of the mercury column in the tube is commonly measured in inches or in millimeters. As you listen to the weather report on the evening news, notice that the atmospheric pressure is given in inches. Where the metric system is in use, millimeters are favored. Millimeters are synonymous with *torr*, a unit named for the Italian scientist Evangalista Torricelli. Less frequently used units are *atmospheres* and *pascals*. The following equation shows the relationship among these various units of pressure.

$$1 \text{ atmosphere} = 14.7 \text{ inches} = 760 \text{ mm} = 760 \text{ torr} = 101{,}325 \text{ pascals}$$

TEMPERATURE

Temperature is another factor influencing the behavior of gases. *Temperature* is a measure of the speed at which molecules and atoms are moving. The higher the temperature, the faster the speed.

Units of Temperature

In place of the familiar Fahrenheit and Celsius (centigrade) units of temperature, the study of gases uses the Kelvin temperature scale, which is based on the idea of absolute zero. The theory is that there is a temperature, called absolute zero (0 K), at which all movement stops. Scientific experimentation has come very close to achieving absolute zero (within thousandths of a degree), but some scientists believe that absolute zero itself is unattainable.

In order to determine how cold absolute zero is, several gases were chilled and graphs were made of their volumes at various temperatures. Each gas became a liquid at a certain low temperature, and from that point on the slope of its graph was extrapolated until it intersected the x axis (temperature). The graph of each of the gases intersected this axis at –273°C (Figure 6.3). From this experimental data it has been concluded that

$$\text{Absolute zero} = 0\text{ K} = -273°\text{C}$$

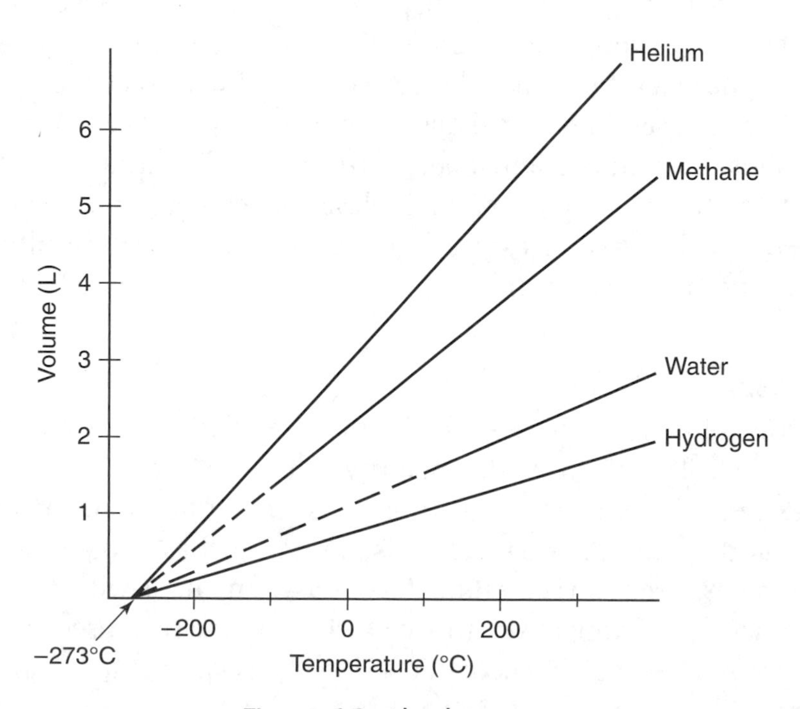

Figure 6.3: Absolute zero

All problems dealing with gases must utilize kelvins (not called degrees Kelvin). To change from degrees Celsius to kelvins, simply add 273. To change from kelvins to degrees Celsius, subtract 273.

EXERCISES

Solve these temperature scale conversion problems.

1. 25°C = ________K **2.** 100°C = ________K

3. 100 K = ________°C **4.** 40 K = ________°C

Answers

1. 298 **2.** 373 **3.** –173 **4.** –233

VOLUME

Volume is the third factor that must be considered in discussing the behavior of gases. *Volume* describes an amount of space. For example, there are 2-liter bottles, 500-cubic-centimeter engines, 65 cubic feet of interior auto space, and 2 tablespoons of butter in a recipe. Volume units include gallons, quarts, liters, cubic centimeters, and ounces. Volume can be either fixed or elastic. A *fixed volume* is contained, as in a soft drink can, and has a definite size. An *elastic volume* is one that can change, like that of a balloon. As you consider problems involving gases, you will need to examine the issue of the flexibility of the volume.

MATHEMATICAL APPROACH TO GASES

The majority of gas behavior can be described mathematically by two equations: the *combined gas law* and the *ideal gas law*. One of these, the combined gas law, is given below and is used for situations involving change. That is, you have information obtained under one set of conditions (P_1, V_1, and T_1) and want to know what will happen under another set of conditions (P_2, V_2, and T_2).

$$\frac{P_1V_1}{T_1} = \frac{P_2V_2}{T_2}$$

In this equation:

P = pressure
V = volume
T = temperature (in kelvins)

The beauty of this equation is that it has widespread applications. It can be used to consider just pressure and volume, just pressure and temperature, just volume and temperature, or all three. You do not have to remember different equations for each of these situations. Use this all-purpose equation keeping only the variables that you need.

Boyle's Law

Robert Boyle was a seventeenth-century Irish scientist who studied the relationship between pressure and volume in gases. He confined his experimentation to these factors, without any change in temperature. As you visualize a situation in which only volume and pressure can change, think of a helium-filled balloon. If you squeeze the balloon to make it smaller, you can feel the pressure inside it become greater and greater. The balloon could even burst because of the pressure. In mathematical terms, **Boyle's Law states that volume and pressure are inversely proportional.** That is, as volume decreases, pressure increases. This law also means that as volume increases, pressure decreases.

Look back at the multipurpose equation above and rewrite it using only the pressure and volume terms—simply omit the temperature terms. What you have written is the mathematical version of Boyle's Law:

$$P_1V_1 = P_2V_2$$

You did not have to learn a new equation but instead used the multipurpose equation you learned earlier and stripped it down to just the pressure and volume terms that you needed.

Here is an example of using the pressure and volume relationship to solve a problem.

Example

How many liters in size will a balloon become at a pressure of 700 mm if it has a volume of 4 liters at a pressure of 600 mm?

$$P_1V_1 = P_2V_2$$

$$(600\text{ mm})(4\text{ liters}) = (700\text{ mm})(V_2)$$

$$\frac{(600)(4)}{700} = V_2$$

$$3.43\text{ liters} = V_2$$

EXERCISES

Answer the following questions using Boyle's Law.

1. How does the popping of bubble wrap illustrate Boyle's Law?
2. If neon gas has a pressure of 200 mm when in a 12-liter tank, what is its pressure when put in a neon sign whose volume is 2 liters?
3. An aerosol can contains 300 milliliters of a compressed gas at a pressure of 4.1 atmospheres. If this gas is sprayed into a plastic bag, what is the volume of the bag if the pressure is 1.0 atmosphere?
4. What pressure (in torr) is necessary to compress 3 liters of carbon dioxide to 2 liters having a pressure of 700 torr?

Answers

1. In popping bubble wrap, the volume of the bubble is made smaller by pushing on it. This increases the internal pressure, causing the bubble to pop.
2. $(200 \text{ mm})(12 \text{ liters}) = (P_2)(2 \text{ liters})$
   $1{,}200 \text{ mm} = P_2$
3. $(4.1 \text{ atm})(300 \text{ ml}) = (1.0 \text{ atm})(V_2)$
   $1{,}230 \text{ ml} = V_2$
4. $(700 \text{ torr})(2 \text{ liters}) = (P_2)(3 \text{ liters})$
   $466.7 \text{ torr} = P_2$

Charles' Law

Jacques Charles, an eighteenth-century French physicist, investigated the relationship between temperature and volume with respect to gases. He was the first to fill a balloon with hydrogen and make a solo hot-air balloon flight. In his experimentation, volume and temperature were allowed to change while pressure was held constant. A volume that is allowed to change brings to mind the flexibility of a balloon. Recalling that molecules move faster as they warm up, it stands to reason that warmer, faster molecules push harder on the walls of their container than colder ones do. Because volume can change here, the volume gets larger as a response to this pushing. In mathematical terms, **Charles' Law states that temperature and volume are directly proportional, or that as temperature increases, volume increases.**

In order to solve problems involving just temperature and volume, remember the multipurpose equation and rewrite it with only the V and T terms:

$$\frac{V_1}{T_1} = \frac{V_2}{T_2}$$

Example

What is the volume of a helium-filled balloon at 200 K if it has a volume of 8 liters at 100 K?

$$\frac{8 \text{ liters}}{100 \text{ K}} = \frac{V_2}{200 \text{ K}}$$

$$(8)(200) = V_2(100)$$

$$1{,}600 = 100V_2$$

$$16 \text{ liters} = V_2$$

EXERCISES

Use Charles' Law to answer the following questions.

1. What temperature (in K) is needed to obtain a volume of 500 liters from a volume of 200 liters at 298 K?
2. What would happen to the size of a balloon put in a freezer?
3. How big will a hot-air balloon become at a temperature of 100°C if it has a volume of 50,000 liters at a temperature of 50°C?
4. If a hot-air balloon has a volume of 300,000 liters at a temperature of 500 K, what is its volume at 300 K?

Answers

1. $$\frac{200 \text{ liters}}{298 \text{ K}} = \frac{500 \text{ liters}}{T_2}$$

   $$745 \text{ K} \quad T_2$$

2. The colder temperature would slow down the molecules of the gas inside the bal-

loon. These slower molecules would not push as hard on the walls of the balloon, allowing it to shrink to a smaller volume.

3. $$\frac{50{,}000 \text{ liters}}{323 \text{ K}} = \frac{V_2}{373 \text{ K}}$$

$$57{,}740 \text{ liters} \quad V_2$$

4. $$\frac{300{,}000 \text{ liters}}{500 \text{ K}} = \frac{V_2}{300 \text{ K}}$$

$$180{,}000 \text{ liters} = V_2$$

Gay-Lussac's Law

From your understanding of how molecules respond to heat and cold, what happens to the pressure of an enclosed gas when heat is applied? As you can imagine, intense pressure builds up as the gas molecules move faster and strike the walls of the container more often and with more force. **Gay-Lussac's Law states that temperature and pressure are directly proportional: As one increases, so does the other.**

Examples

1. Rewrite the all-purpose equation used throughout this chapter to include only the pressure and temperature terms.

$$\frac{P_1}{T_1} = \frac{P_2}{T_2}$$

2. Calculate the pressure that results when the temperature becomes 100°C if the pressure at 25°C is 750 torr.

$$\frac{P_1}{T_1} = \frac{P_2}{T_2}$$

$$\frac{750}{298} = \frac{P_2}{373}$$

$$P_2 = \frac{(750)(373)}{298} = 938.8 \text{ torr}$$

EXERCISES

Answer the following questions dealing with pressure and temperature.

1. Why is there a warning label on aerosol cans that cautions the user against incinerating the can?
2. If the pressure inside a can is 950 mm when the temperature is 330 K, what is the pressure when the can is chilled to 273 K?
3. In order to lower the pressure in an oxygen tank to 500 torr, what should its temperature be if the pressure is 800 torr at 400 K?
4. The air in a scuba tank is at 40°C. What is its pressure in atmospheres if the tank's pressure is 2.4 atmospheres at 80°C?

Answers

1. Incinerating the can causes the temperature of the contents to increase, therefore increasing the internal pressure. The can might be unable to withstand the pressure increase and could explode.

2. $$\frac{950\text{ mm}}{330\text{ K}} = \frac{P_2}{273\text{ K}}$$
$$P_2 = 785.90\text{ mm}$$

3. $$\frac{800\text{ torr}}{400\text{ K}} = \frac{500\text{ torr}}{T_2}$$
$$250\text{ K} = T_2$$

4. $$\frac{2.4\text{ atm}}{353\text{ K}} = \frac{P_2}{313\text{ K}}$$
$$2.1\text{ atm} = P_2$$

COMBINATION THEORY

As you can imagine, most real applications of gas theory involve varying the temperature, pressure, and volume. This is the time to use the entire multipurpose equation that we began with.

Example

What pressure is expected to develop at a volume of 10 liters and a temperature of 25°C if the volume is 5 liters, the temperature is 50°C, and the pressure is 30 inches of mercury?

$$\frac{(30\text{ inches})(5\text{ liters})}{50+273} = \frac{(P_2)(10\text{ liters})}{25+273}$$

$$\frac{(30)(5)}{323} = \frac{(10)(P_2)}{298}$$

$$(30)(5)(298) = (10)(P_2)(323)$$

$$\frac{(30)(5)(298)}{(10)(323)} = P_2$$

$$13.84\text{ inches} = P_2$$

EXERCISES

Use combination theory to solve these problems.

1. Calculate the volume that develops at a pressure of 600 torr and 200 K if there is a volume of 2.2 liters at 700 torr and 100 K.
2. How many atmospheres of oxygen are in a 48-liter tank at 300 K if that same amount of oxygen occupies 45 liters of space at a pressure of 1.2 atmospheres and a temperature of 200 K?
3. What is the temperature of 400 liters of nitrogen at a pressure of 29.8 inches if it has a temperature of 25°C at a pressure of 30.1 inches and a volume of 500 liters?

Answers

1. $$\frac{(700\text{ mm})(2.2\text{ liters})}{100\text{ K}} = \frac{(600\text{ mm})(V_2)}{200\text{ K}}$$

$$5.13\text{ liters} = V_2$$

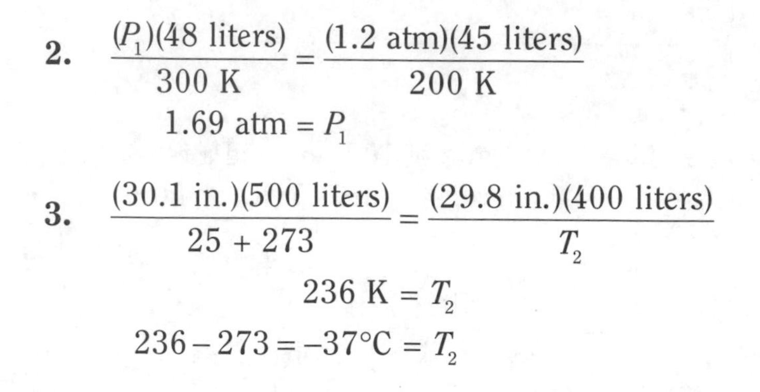

2. $$\frac{(P_1)(48 \text{ liters})}{300 \text{ K}} = \frac{(1.2 \text{ atm})(45 \text{ liters})}{200 \text{ K}}$$
$$1.69 \text{ atm} = P_1$$

3. $$\frac{(30.1 \text{ in.})(500 \text{ liters})}{25 + 273} = \frac{(29.8 \text{ in.})(400 \text{ liters})}{T_2}$$
$$236 \text{ K} = T_2$$
$$236 - 273 = -37°\text{C} = T_2$$

STP

In chemistry the term *STP* does not refer to an oil additive for cars but rather is an acronym for *standard temperature and pressure*. Standard temperature has been set by the scientific community to be 0°C, and standard pressure to be 1 atmosphere or 760 mm or 760 torr. Because gas experiments are performed under a variety of temperature and pressure conditions in laboratories all over the world, it is necessary that scientists have some type of standard so that they can compare results. STP is this standard, and results are often said to be "corrected to STP." That is, the results have been calculated to be the values that they would be at STP.

Example

Suppose you perform an experiment in your lab and have 500 ml of oxygen at 25°C and 750 mm of pressure. What is the volume if the experiment is conducted at STP?

$$\frac{P_1V_1}{T_1} = \frac{P_2V_2}{T_2}$$

$$\frac{(750 \text{ mm})(500 \text{ ml})}{273 + 25} = \frac{(760 \text{ mm})(V_2)}{273} \qquad (P_2 \text{ and } V_2 \text{ are from STP})$$

$$452 = V_2$$

EXERCISES

Solve these STP problems.

1. If you have 5 liters of nitrous oxide at STP, what is its volume under room conditions of 748 mm pressure and 18°C?

2. How large (in liters) is a helium-filled balloon at STP if it has a volume of 5.5 liters at 10°C and 0.95 atmosphere of pressure?

Answers

1. $$\frac{(5 \text{ liters})(760 \text{ mm})}{273 \text{ K}} = \frac{(748 \text{ mm})(V_2)}{18+273} \qquad V_2 = 5.42 \text{ liters}$$

2. $$\frac{(5.5 \text{ liters})(0.95 \text{ atm})}{283 \text{ K}} = \frac{V_2(1 \text{ atm})}{273} \qquad V_2 = 5.04 \text{ liters}$$

IDEAL GAS LAW

Chemistry students everywhere say "PIV NERT," but the ideal gas law is really

$$PV = nRT$$

where

P = pressure
V = volume (in liters)
n = number of moles of the gas
R = a constant whose value is 0.0821 if the pressure is in atmospheres, or 62.3 if the pressure is in millimeters or torr
T = temperature (in K)

Example

Ideal gas law
What pressure (in millimeters), is expected from 0.55 mole of nitrogen that is at 20°C and in a 0.75-liter container?

$$PV = nRT$$
$$P(0.75 \text{ liter}) = 0.55 \text{ mole}(62.3)(293 \text{ K})$$
$$P = 13{,}386.2 \text{ millimeters or torr}$$

Because the ideal gas law has four variables (P, V, n, and T), you would expect to see problems in which three of these variables would be given and the fourth the unknown. If your algebra skills for dealing with a variable in different positions in an equation need some brushing up, turn to Unit 15 and work through the algebra section.

EXERCISES

Solve the following problems dealing with the ideal gas law.

1. How many moles cause a pressure of 500 torr in a 2.5-liter container at 50°C?
2. If there is 0.2 mole of sulfur dioxide gas in a 5.0-liter tank kept at 20°C, what is the pressure in the tank? Calculate this pressure in millimeters.
3. What is the temperature (in kelvins) of 0.3 mole of nitrogen stored in a 12-liter container at a pressure of 0.5 atmosphere?

Answers

1. $(500 \text{ torr})(2.5 \text{ liters}) = n(62.3)(323 \text{ K})$
   $n = 0.06$ mole
2. $(P)(5.0 \text{ liters}) = (0.2 \text{ mole})(62.3)(293 \text{ K})$
   $P = 730.2$ mm
3. $(0.5 \text{ atm})(12 \text{ liters}) = (0.3 \text{ mole})(0.0821)(T)$
   $T = 243.6$ K

The ideal gas law can be used as a gateway to finding molecular weight. Recall that

$$\text{Molecular weight} = \frac{\text{grams}}{\text{mole}}$$

When you know the number of grams and can use the ideal gas law to calculate moles, you have the necessary values for finding molecular weight.

Example

Calculate the molecular weight of an unknown gas if 0.35 gram of it occupies a volume of 0.1 liter at a pressure of 700 mm and a temperature of 298 K.

Step 1. Calculate moles

$$(700 \text{ mm})(0.1 \text{ liter}) = n(62.3)(298 \text{ K})$$
$$0.0038 = n$$

Step 2. Calculate molecular weight using $\frac{\text{grams}}{\text{mole}}$

$$\text{Molecular weight} = \frac{0.35 \text{ gram}}{0.0038 \text{ mole}} = 92 \text{ grams/mole}$$

EXERCISES

Solve the following problems dealing with the ideal gas law.

1. Calculate the molecular weight of 1.15 grams of a gas if 3.90 grams of it exerts a pressure of 1.3 atmospheres inside a 400-milliliter flask at 300 K.

2. What is the molecular weight of a gas if 3.90 grams of it exerts a 740-mm pressure inside a 1.5-liter flask at 350 K?

Answers

1. $PV = nRT$
$(1.3 \text{ atm})(0.400 \text{ liter}) = n(0.0821)(300)$
$n = 0.02$ mole

$$\text{Molecular weight} = \frac{1.15 \text{ grams}}{0.02 \text{ mole}} = 57.5 \text{ grams/mole}$$

2. $PV = nRT$
$(740 \text{ mm})(1.5 \text{ liters}) = n(62.3)(350)$
$n = 0.05$ mole

$$\text{Molecular weight} = \frac{3.90 \text{ grams}}{0.05 \text{ mole}} = 78 \text{ grams/mole}$$

We have now covered the primary gas topics. There is, however, another concept that will round out your study of gases.

DALTON'S LAW OF PARTIAL PRESSURES

John Dalton, an English scientist, studied the effect of more than one gas in a closed environment. His law of partial pressures states that the **total pressure for a mixture of**

gases in a container is the sum of the partial pressures of the gases present. The partial pressure of a gas is the pressure that the gas would exert if it were alone in the container. In mathematical terms:

$$P_T = P_A + P_B + P_C + \cdots$$

where

P_T = total pressure
P_A = partial pressure of gas A
P_B = partial pressure of gas B
P_C = partial pressure of gas C

The most common use of this law has to do with the practice of collecting a gas by allowing it to bubble through water. Naturally the gas picks up some water, and so the total pressure reflects the pressure of the gas itself plus the pressure of the water vapor mixed with it.

Example

Hydrogen gas is collected over water at 25°C. At this temperature the vapor pressure of water is 24 mm. The total pressure of the hydrogen and the water vapor is 740 mm. What is the pressure of just the hydrogen?

$$P_T = P_{water} + P_{hydrogen}$$
$$740 = 24 + P_{hydrogen}$$
$$716 = P_{hydrogen}$$

EXERCISE

Answer the following question dealing with the law of partial pressures.

1. What is the pressure of oxygen collected over water at 17°C if the total pressure of the system is 735 torr and the vapor pressure of water at this temperature is 15 torr?

Answer

1. $735 = 15 + P_{oxygen}$

   $720 \text{ torr} = P_{oxygen}$

GAS REVIEW SET EXERCISES

The following exercises address all the topics presented in this unit. They are not grouped by type, nor are they in any particular order.

1. What is the size of a balloon in liters at 20°C and 30 inches of pressure if it has a volume of 12 liters at 30°C and 29 inches of pressure?
2. Calculate the temperature in kelvins for 3 liters of nitrogen at a pressure of 600 mm if there is 0.2 mole of nitrogen present.
3. What is the volume of helium at 1.2 atmospheres if the volume is 3 liters at a pressure of 0.8 atmosphere?
4. How many moles of oxygen are present in a 5-liter tank at 0°C at a pressure of 750 torr?
5. What is the volume of hydrogen if 0.5 mole of it at 100°C exerts a pressure of 800 mm?
6. What is the temperature of a gas at a pressure of 30.1 inches of mercury and a volume of 40 liters if at 30.8 inches of pressure it has a volume of 45 liters and a temperature of 25°C?
7. The gas in a 15-liter tank at 28°C is put into a 5-liter container with no change in pressure. What should the temperature be (in degrees) Celsius?
8. What is the pressure (in atmospheres) of 505 milliliters of a gas at 70°C if there is 0.9 mole present?
9. Suppose that a gas/air mixture in a 0.75-liter automobile cylinder at 1 atmosphere is compressed to 0.07 liter by a piston. What is the resulting pressure?
10. If the pressure of the gas in a can of hair spray at a room temperature of 300 kelvins is 2 atmospheres, what is the pressure when the can is heated to 450 kelvins?
11. Scuba divers use a mixture of helium and oxygen in their tanks. What is the pressure of the oxygen in a tank if the pressure caused by the helium is 200 torr and the total pressure in the tank is 925 torr?
12. Calculate the molecular weight of mystery gas *X* if 2.4 grams of it occupies a volume of 0.75 liter at a temperature of 303 kelvins with a pressure of 800 torr.

13. What is the pressure (in mm Hg), of 0.15 mole of methane gas in an 8-liter tank at 400 K?

14. If a weather balloon has a volume of 950 liters at a room temperature of 25°C, what is its volume at 0°C?

15. If a hot-air balloon has a volume of 35,000 liters at a temperature of 20°C when it is tethered to the ground, what is its volume when the burners are turned on and the temperature rises to 120°C?

16. If an elastic packing balloon has a volume of 3.2 liters when its pressure is 700 mm, what is its volume when the pressure is decreased to 600 mm?

17. Calculate the volume of nitrogen expected if 0.5 mole of it is confined in a container at 300 K at a pressure of 1.5 atmospheres.

18. What is the molecular weight of a gas if 1.18 grams of it creates a pressure of 550 torr in a 1-liter container at 25°C?

19. The pressure in an aerosol can of furniture polish is 2.1 atmospheres at a room temperature of 24°C. What is the temperature (in degrees Celsius) when the can is thrown into an incinerator where the temperature increases to 300°C?

20. In an experiment done at the high altitudes of the Andes, the data obtained was 3.15 liters of a mystery gas at a temperature of 18°C and a pressure of 0.75 atmosphere. What is its volume at sea level with a pressure of 1.0 atmosphere and a temperature of 22°C?

21. Carbon dioxide gas collected in the lab had a volume of 5 liters at 70°C and 600 mm pressure. What is its volume at STP?

22. If the measurements of a weather balloon are 150 liters of volume at a temperature of 37°C and a pressure of 770 mm, what is its volume at STP?

Answers

1.

$$\frac{P_1V_1}{T_1} = \frac{P_2V_2}{T_2}$$

$$\frac{(29)(12)}{303} = \frac{(30)(V_2)}{293}$$

$$11.22 = V_2$$

2.

$$PV = nRT$$

$$(600)(3) = 0.2(62.3)(T)$$

$$144.46 \text{ K} = T$$

3. $P_1V_1 = P_2V_2$
$(0.8)(3) = (1.2)V_2$
$2 \text{ liters} = V_2$

4. $PV = nRT$
$(750)(5) = n(62.3)(273)$
$0.22 = n$

5. $PV = nRT$
$(800)(V) = 0.5(62.3)(373)$
$V = 14.52 \text{ liters}$

6. $$\frac{P_1V_1}{T_1} = \frac{P_2V_2}{T_2}$$
$$\frac{(30.8)(45)}{298} = \frac{(30.1)(40)}{T_2}$$
$$258.87 \text{ K} = T_2$$

7. $$\frac{V_1}{T_1} = \frac{V_2}{T_2}$$
$$\frac{15 \text{ liters}}{301 \text{ K}} = \frac{5 \text{ liters}}{T_2}$$
$$100.33 \text{ K} = T_2$$

8. $PV = nRT$
$P(0.505) = 0.9(0.0821)(343)$
$P = 50.19 \text{ atmospheres}$

9. $P_1V_1 = P_2V_2$
$(1 \text{ atm})(0.75 \text{ liter}) = P_2(0.07 \text{ liter})$
$10.71 \text{ atm} = P_2$

10. $$\frac{P_1}{T_1} = \frac{P_2}{T_2}$$
$$\frac{2 \text{ atm}}{300 \text{ K}} = \frac{P_2}{450 \text{ K}}$$
$$3 \text{ atm} = P_2$$

11. $P_{total} = P_{helium} + P_{oxygen}$
$925 = 200 + P_{oxygen}$
$725 = P_{oxygen}$

12. Step 1: Solve $PV = nRT$ for moles (n).

$$(800 \text{ torr})(0.75 \text{ liter}) = n\,(62.3)(303 \text{ K})$$
$$0.032 = n$$

Step 2: molar mass = grams/mole

$$= 2.4 \text{ g}/0.032 \text{ mole}$$
$$= 75 \text{ grams/mole}$$

13.
$$PV = nRT$$
$$P(8) = 0.15(62.3)(400)$$
$$P = 467.25 \text{ mm}$$

14.
$$\frac{V_1}{T_1} = \frac{V_2}{T_2}$$
$$\frac{950 \text{ liters}}{298 \text{ K}} = \frac{V_2}{273 \text{ K}}$$
$$870.30 = V_2$$

15.
$$\frac{V_1}{T_1} = \frac{V_2}{T_2}$$
$$\frac{35{,}000}{20 + 273} = \frac{x \text{ liters}}{120 + 273}$$
$$293x = 13{,}755{,}000$$
$$x = 46{,}945.4 \text{ liters}$$

16.
$$P_1V_1 = P_2V_2$$
$$(700)(3.2) = (600)V_2$$
$$3.14 \text{ liters} = V_2$$

17.
$$PV = nRT$$
$$(1.5)V = 0.5(0.0821)(300)$$
$$V = 8.2 \text{ liters}$$

18.
$$PV = nRT$$
$$500(1) = n(62.3)(25 + 273)$$
$$0.027 = n$$
$$\text{molar mass} = \frac{\text{grams}}{\text{mole}}$$
$$= \frac{1.18 \text{ grams}}{0.027 \text{ mole}}$$
$$= 43.7 \text{ grams/mole}$$

19.
$$\frac{P_1}{T_1} = \frac{P_2}{T_2}$$
$$\frac{2.1}{24+273} = \frac{x}{300+273}$$
$$297x = 1{,}203.3$$
$$x = 4.05 \text{ atmospheres}$$

20.
$$\frac{P_1V_1}{T_1} = \frac{P_2V_2}{T_2}$$
$$\frac{0.75(3.15)}{18+273} = \frac{1.0x}{22+273}$$
$$291x = 696.9$$
$$x = 2.4 \text{ liters}$$

21. STP values are 760 mm of pressure and 0°C.
$$\frac{P_1V_1}{T_1} = \frac{P_2V_2}{T_2}$$
$$\frac{(600)(5)}{70+273} = \frac{(760)V_2}{0+273}$$
$$(260{,}680)V = 819{,}000$$
$$V = 3.14 \text{ liters}$$

UNIT 7

Solution Chemistry

Solutions, Molarity, Molality, Percent Solutions, and Colligative Properties

Solutions are involved in all aspects of our lives, from the tea and soft drinks that accompany our meals to the liquid detergents used to wash dishes, the antifreeze solution that protects the functioning of automobiles, and solutions that drip through intravenous lines in hospitals.

Solutions are combinations of at least two ingredients, such as water and salt. The substance present in the larger quantity is the *solvent*, and the one present in the smaller quantity is the *solute*. The hallmark of solutions is their uniformity. Every part of a solution is the same as any other part. This is called *homogeneity*. **A solution is a homogeneous mixture.** For example, when cola is examined microscopically, the same number of sugar molecules are present in each ounce. For most solutions, the solvent is water, although there are certainly numerous exceptions.

Solutions do not have to be liquids. Air is a solution composed of nitrogen, oxygen, water vapor, carbon dioxide, and other gases. The nitrogen is the solvent, as it accounts for nearly 80 percent of any sample of air. There are even solid solutions. Perhaps the most common is brass, a solution of copper and zinc.

A most important aspect of solutions is their concentration. In general, the higher the ratio of solute to solvent, the more concentrated the solution. Kool-Aid is a drink that can be prepared to be as dilute or as concentrated as individual preference dictates. The addition of water decreases the concentration, or in other words, makes the Kool-Aid more dilute. Many solution concentrations are prescribed by their use. A chain saw requires a different concentration of two-cycle engine oil in gasoline than a string trimmer does. The wrong concentration can cause the engine to run inefficiently or even not at all. Eye drops suitable for human use can be harmful to canine eyes because of different concentration requirements.

SATURATION

When the solute and solvent are put together, there is a limit as to how much solute can dissolve. When sugar is added to iced tea and all of it dissolves, the solution is said to be *unsaturated.* When too much sugar is stirred into iced tea, some of the sugar does not dissolve and can be seen collecting at the bottom of the glass. The liquid in the glass has dissolved as much sugar as it can, and so the solution is said to be *saturated*. It is possible under some circumstances to have more solute dissolved than is theoretically possible. This is an unstable and temporary situation called *supersaturation.* Solubility tables list chemicals and their solubilities in grams per 100 ml and allow chemists to plan the preparation of solutions to conform to what is possible.

MOLARITY

In chemistry the most common concept involving concentration is *molarity,* which is abbreviated M (be certain to use a capital M because a lowercase m has a different meaning). **Molarity is the number of moles of solute per liter of solution.**

$$M = \frac{\text{moles}}{\text{liter}}$$

Notice that this is not the same as saying "per liter of water." This distinction is accomplished in the laboratory by using a volumetric flask (Figure 7.1), which is specially designed for solution preparation. Volumetric flasks are manufactured in a variety of sizes, and so the preparer chooses the flask to correspond to the volume of solution desired. First, the solute is added to the flask. Then, water or another solvent is added until the liquid level rises to the etched line on the neck of the flask.

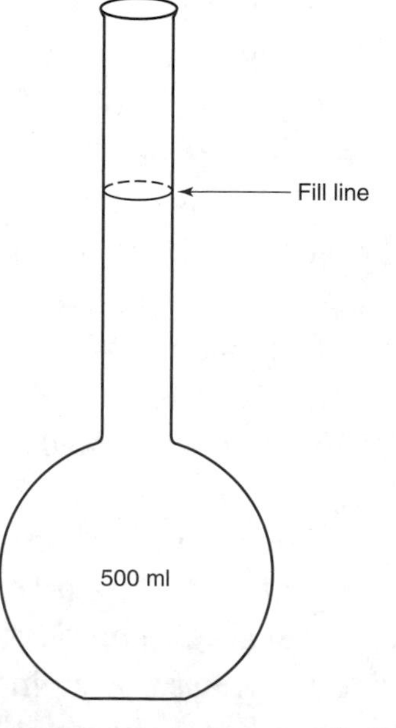

Figure 7.1: Volumetric flask

When solving molarity problems, remember the following.

- Always start with the equation M = moles/liter of solution.
- Plug into this equation whatever information the problem gives you.
- Put an x wherever you have no number.
- Solve for x.
- Ask yourself, "Are we there yet?" Is x what the problem asks for?
- If not, transform x into what is asked for.
- If you have any trouble with the math, review the algebra section in Unit 15.

Examples

Molarity problems

Note: Looking at the molarity equation M = moles/liter, you can see that x can be in one of three locations: M, moles, or liter.

1. What is the molarity of a solution made by dissolving 0.3 mole of $Ca(NO_3)_2$ in enough water to make 500 ml of solution?

$$\begin{aligned} M &= \frac{\text{moles}}{\text{liter}} \\ &= \frac{0.3 \text{ mole}}{0.5 \text{ liter}} \\ &= 0.6 \text{ mole/liter} \end{aligned}$$

2. How much 0.5 M NaCl can be made from 0.1 mole of NaCl?

$$\begin{aligned} M &= \frac{\text{moles}}{\text{liter}} \\ 0.5 \text{ M} &= \frac{0.1 \text{ mole}}{x \text{ liters}} \\ 0.5x &= 0.1 \\ x &= 0.2 \text{ liter} \end{aligned}$$

3. What is the molarity of an epsom salts ($MgSO_4$) solution made by dissolving 6 grams of it in enough water to make 2 liters of solution?

$$M = \frac{\text{moles}}{\text{liter}}$$

Make a slight detour to change grams to moles:

$$6 \text{ grams } MgSO_4 \times \frac{1 \text{ mole } MgSO_4}{120 \text{ grams}} = 0.05 \text{ mole}$$

$$M = \frac{0.05 \text{ mole}}{2 \text{ liters}} = 0.025 \text{ mole/liter}$$

4. How many grams of KOH are needed in order to prepare 2 liters of a 0.25 M solution?

$$M = \frac{\text{moles}}{\text{liter}}$$

$$0.25 \text{ M} = \frac{x \text{ moles}}{2 \text{ liters}}$$

$$0.5 = x \text{ moles}$$

But the question is "How many grams of KOH," so you will need to change moles to grams:

$$0.5 \text{ mole KOH} \times \frac{56 \text{ grams KOH}}{1 \text{ mole KOH}} = 28 \text{ grams of KOH}$$

EXERCISES

Solve the following molarity problems.

1. How many moles of NaCl are needed in order to prepare 750 ml of a 0.2 M solution?
2. What is the molarity of a KNO_3 solution made by dissolving 5.05 grams in enough water to make 2 liters of solution?
3. How many grams of NaOH are needed to make 500 ml of a 0.1 M solution?
4. How many liters of a 0.1 M sugar ($C_{12}H_{22}O_{11}$) solution can be made from 0.5 mole of sugar?
5. How many moles of battery acid (H_2SO_4) are there in 4 liters of 0.25 M solution?
6. What is the molarity of an antifreeze solution that has 50 grams of antifreeze ($C_2H_6O_2$) dissolved in 3 liters of water?
7. How many grams of epsom salts ($MgSO_4$) should be used to make 800 ml of an 0.5 M solution?

8. A generic cleaning solution has a molarity of 0.75. How many liters of it are needed in order to have 0.15 mole?

Answers

1.
$$M = \frac{\text{moles NaCl}}{\text{liter}}$$
$$0.2\text{ M} = \frac{x\text{ moles NaCl}}{0.75\text{ liter}}$$
$$0.15 = \text{moles NaCl}$$

2. Calculate moles from the number of grams given in order to use the molarity equation.

$$5.05\text{ grams KNO}_3 \times \frac{1\text{ mole KNO}_3}{101\text{ grams}} = 0.05\text{ mole KNO}_3$$
$$M = \frac{\text{moles KNO}_3}{\text{liter}}$$
$$= \frac{0.05\text{ mole KNO}_3}{2\text{ liters}}$$
$$= 0.025\text{ mole/liter}$$

3.
$$M = \frac{\text{moles NaOH}}{\text{liter}}$$
$$0.25\text{ M} = \frac{x\text{ moles NaOH}}{0.5\text{ liter}}$$
$$= 0.125\text{ moles NaOH}$$

4.
$$M = \frac{\text{moles sugar}}{\text{liter}}$$
$$0.1\text{ M} = \frac{0.5\text{ mole sugar}}{x\text{ liters}}$$
$$5\text{ liters} = x$$

5.
$$0.25 = \frac{x\text{ moles}}{4\text{ liters}}$$
$$(4)(0.25) = 1\text{ mole}$$

6. Find the number of moles first:

$$50 \text{ g antifreeze} \times \frac{1 \text{ mole}}{62 \text{ g}} = 0.81 \text{ mole}$$

$$x \text{ M} = \frac{0.81 \text{ mole}}{3 \text{ liters}} = 0.27 \text{ M}$$

7.

$$0.5 \text{ M} = \frac{x \text{ moles}}{0.8 \text{ liter}} = 0.4 \text{ mole}$$

$$0.4 \text{ mole} \times \frac{120 \text{ grams}}{1 \text{ mole}} = 48 \text{ grams}$$

8.

$$0.75 \text{ M} = \frac{0.15 \text{ mole}}{x \text{ liters}}$$

$$0.75x = 0.15$$

$$x = 0.15/0.75 = 20 \text{ liters}$$

MOLALITY

A less frequently used measure of solution concentration is molality, abbreviated as a lowercase italic *m*. **Molality is the number of moles of solute per kilogram of solvent.** Notice that the numerator in molality calculations is the same as the numerator in molarity calculations, but that the denominators are different. For molality, the denominator differs in two respects: It is in kilograms rather than liters and it involves solvent rather than solution. For the preparation of molal solutions, a volumetric flask is not needed. This is a preparation based only on weight. Molality is expressed in moles/kilogram.

$$\text{molality} = \frac{\text{moles of solute}}{\text{kilograms of solvent}}$$

Examples

Molality

1. Calculate the molality of an antifreeze solution made by dissolving 186 grams of antifreeze ($C_2H_6O_2$) in 2 kilograms of water.

   Calculate the moles of antifreeze:

$$186 \text{ grams antifreeze} \times \frac{1 \text{ mole antifreeze}}{62 \text{ grams}} = 3 \text{ moles antifreeze}$$

$$m = \frac{3 \text{ moles antifreeze}}{2 \text{ kilograms}} = 1.5 \text{ moles/kilogram}$$

2. How many grams of lye (NaOH) should be added to 800 grams of water in order to make a 0.5 m solution? Remember that the denominator must be in kilograms.

$$0.5\ m = \frac{x \text{ moles NaOH}}{0.8 \text{ kilogram}}$$

$$0.4 = \text{moles NaOH}$$

Because the question asks for grams, the 0.4 mole of NaOH must be changed to grams.

$$0.4 \text{ mole NaOH} \times \frac{40 \text{ grams NaOH}}{1 \text{ mole}} = 16 \text{ grams NaOH}$$

EXERCISES

Solve the following molality problems.

1. Calculate the molality of a solution of calcium chloride ($CaCl_2$) made by dissolving 222 grams in 1 kilogram of water.
2. How many grams of sodium fluoride (NaF) must be mixed with 500 grams of water in order to make a 0.1 m solution?
3. For a 0.5 molal solution of Drano (NaOH) how many grams of Drano should be added to 1.5 kilograms of water?
4. What is the molality of a salt (NaCl) solution that has 0.15 mole dissolved in 3.00 kilograms of water?

Answers

1. $222 \text{ grams } CaCl_2 \times \frac{1 \text{ mole } CaCl_2}{111 \text{ grams}} = 2 \text{ moles } CaCl_2$

$$m = \frac{\text{moles } CaCl_2}{\text{kilogram water}}$$

$$= \frac{2 \text{ moles } CaCl_2}{1 \text{ kilogram water}}$$

$$= 2 \text{ moles/kilogram}$$

2. $0.1\ m = \dfrac{x \text{ moles NaF}}{0.5 \text{ kg}}$

$0.5 = \text{moles NaF}$

The problem asks for grams of NaF, therefore, moles must be converted to grams.

$$0.05 \text{ mole NaF} \times \frac{42 \text{ grams NaF}}{1 \text{ mole}} = 2.1 \text{ grams NaF}$$

3. $0.5\ m = \dfrac{x \text{ moles}}{1.5 \text{ kg}}$

$0.5(1.5) = 0.75 \text{ mole}$

$$0.75 \text{ mole} \times \frac{40 \text{ grams}}{1 \text{ mole}} = 30 \text{ grams NaOH}$$

4. $m = 0.15 \text{ mole}$

3 liters = 0.05

PERCENT SOLUTIONS

Whereas molarity and molality are the concentration units of choice for chemists, percent solutions are more suitable for consumer products. The three types of percent solutions commonly used are defined as follows.

1. Weight/weight solutions:

$$\frac{\text{grams of solute}}{\text{grams of solution}} \times 100 = \% \text{ solution}$$

For example, mixing 5 grams of solute with 95 grams of solvent produces 100 grams of solution.

$$\frac{5 \text{ grams of solute}}{100 \text{ grams solution}} \times 100 = 5\% \text{ solution}$$

2. Weight/volume solutions:

$$\frac{\text{grams of solute}}{\text{milliliters of solution}} \times 100 = \% \text{ solution}$$

For example, mixing 2 grams of solute with enough water to make 50 milliliters of solution:

$$\frac{\text{2 grams of solute}}{\text{50 ml of solution}} \times 100 = 4\% \text{ solution}$$

3. Volume/volume solutions:

$$\frac{\text{milliliters of solute}}{\text{milliliters of solution}} \times 100 = \% \text{ solution}$$

For example, mixing 5 milliliters of alcohol with enough water to make 100 milliliters of solution:

$$\frac{\text{5 milliliters of solute}}{\text{100 milliliters of solution}} \times 100 = 5\% \text{ solution}$$

Note: In a volume/volume percent solution, you cannot measure out 95 milliliters of solvent and think that with 5 milliliters of solute it will equal 100 milliliters of solution. This would require the use of a volumetric flask. Adding some liquids together results in a lower volume than predicted because of spaces between molecules.

For the most part, the label on consumer products does not indicate which of these percent formulas is used. Typically, a volume/volume calculation is used for solutions made from two liquids. Each manufacturer chooses the method of calculating percent that is most suitable for a particular product. An urban legend of yesteryear claimed that the beer in one state had more alcohol than the beer in an adjacent state. The truth was that one state used a weight/volume percent calculation, whereas the other used a volume/volume approach, and the beers were essentially identical in alcohol content.

EXERCISES

Solve these percent solution problems.

1. Calculate the weight/volume percent solution formed by adding 10 grams of salt (NaCl) to enough water to make 500 milliliters of solution.
2. Calculate the volume/volume percent solution formed by combining 25 milliliters of ethanol with enough water to make 200 milliliters of solution.
3. Calculate the weight/weight percent solution formed by mixing 10 grams of sugar with 90 grams of water.

4. If a sugar solution is made by adding 30 grams of sugar to enough water to make 600 milliliters of solution, what percent will it be? What kind of percent problem is this—weight/weight, weight/volume, or volume/volume?
5. What is the percent of salt in seawater if a 150-gram sample is found to have 5.25 grams of salt itself?
6. If a 20-milliliter dose of cough syrup is labeled as having 4.2 percent of the active ingredient, how many milliliters of active ingredient are present in the dose?

Answers

1. $$\frac{10 \text{ grams salt}}{500 \text{ milliliters solution}} \times 100 = 2\%$$

2. $$\frac{25 \text{ milliliters ethanol}}{200 \text{ milliliters solution}} \times 100 = 12.5\%$$

3. $$\frac{10 \text{ grams sugar}}{100 \text{ grams solution}} \times 100 = 10\%$$

4. $$\frac{30 \text{ grams of sugar}}{600 \text{ ml solution}} \times 100 = 5\%$$

5. $$\frac{5.25 \text{ grams of salt}}{150 \text{ grams seawater}} \times 100 = 3.5\%$$

6. $$\frac{x \text{ ml of active ingredient}}{20 \text{ ml cough syrup}} \times 100 = 4.2\%$$

OTHER CONCENTRATION TERMS

The remaining concentration terms are parts per million (ppm) and parts per billion (ppb). These terms are usually reserved for describing very small amounts, such as levels of air and water pollutants.

Parts per million is the number of solute parts in every million solution parts. The reason the word *parts* is used is that parts per million is a ratio in which the parts can be any unit.

$$\text{ppm} = \frac{\text{parts solute}}{1{,}000{,}000 \text{ parts solution}}$$

Parts per billion is the number of solute parts in every billion solution parts.

$$\text{ppb} = \frac{\text{parts solute}}{1{,}000{,}000{,}000 \text{ parts solution}}$$

COLLIGATIVE PROPERTIES

Colligative properties are properties of solutions that are related to the number of particles of solute rather than to the chemical identity of these solute particles. These solution properties include vapor pressure, boiling point, and freezing point.

Vapor Pressure

Vapor pressure is caused by the evaporation of molecules at the surface of a liquid. These escaping molecules exert an upward pressure as they leave the liquid (Figure 7.2). For example, gasoline has a greater vapor pressure than syrup because molecules of gasoline evaporate more readily than molecules of syrup. Perfume and alcohol have vapor pressures that are greater than the vapor pressure of water.

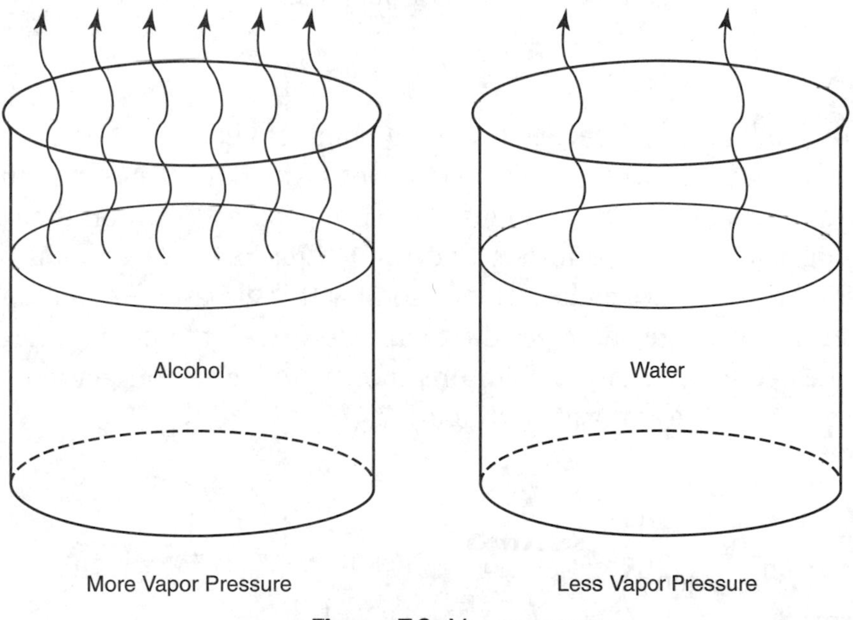

Figure 7.2: Vapor pressure

The vapor pressure of a liquid is reduced by the presence of solute particles in the liquid because the solute molecules reduce the number of the liquid molecules on the surface. It is the surface molecules that evaporate and therefore create vapor pressure. The vapor pressure of water is greater than the vapor pressure of a sugar solution because the presence of sugar molecules at the liquid's surface decreases the number of water molecules that are free to evaporate (Figure 7.3).

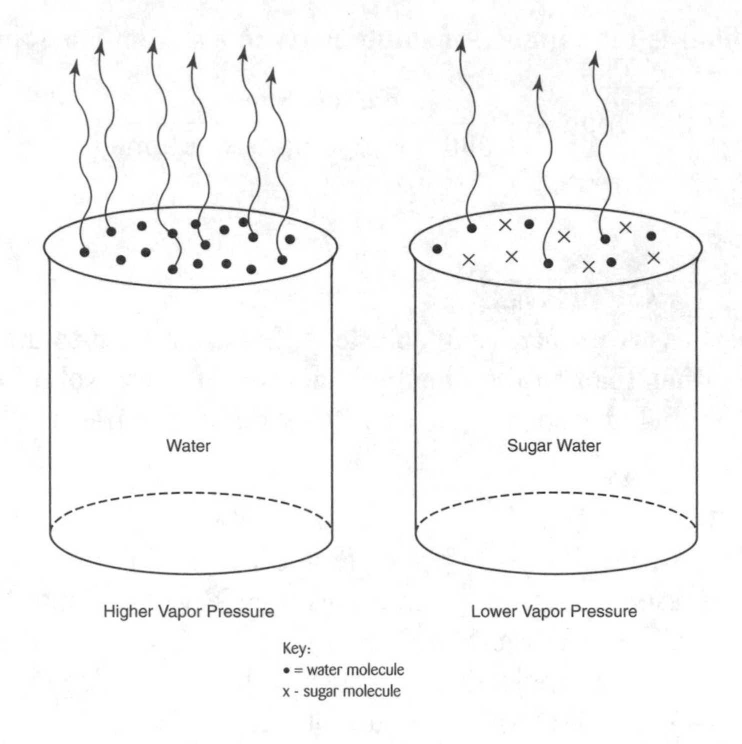

Figure 7.3: Solute and vapor pressure

Boiling Point

The second of the colligative properties is boiling point. Boiling point and vapor pressure are related. The *boiling point* is the temperature at which the vapor pressure escaping from a liquid is equal to the atmospheric pressure pushing down on the surface of the liquid. **Boiling point elevation means that the boiling point of a liquid increases when molecules of a solute are added.** The reason for this elevation in boiling point is that the presence of the solute decreases the vapor pressure of the liquid; therefore the liquid has to be heated to a higher temperature in order to have enough vapor pressure to match the pressure of the atmosphere (Figure 7.4).

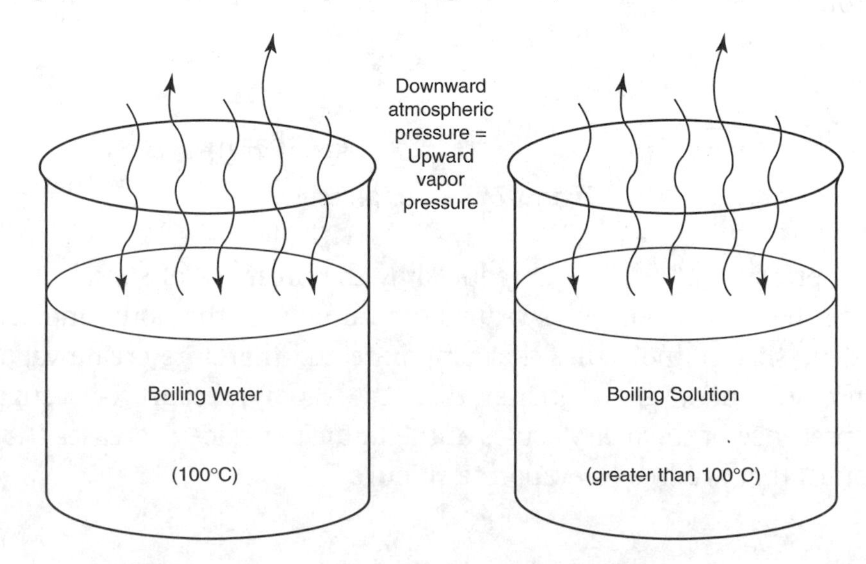

Figure 7.4: Boiling point diagram

EXERCISES

Answer the following questions dealing with boiling point.

1. Would water boil at a cooler or a hotter temperature on Mount Everest compared with a location at sea level?
2. What can you say about the temperatures of two pots of boiling water, one with salt added and one without?
3. If the antifreeze solution in a car's radiator boils (not a desirable thing), is its temperature higher, lower, or the same compared to that of plain water boiling on a stove?

Answers

1. On Mount Everest the atmospheric pressure is extremely low, and so the water does not have to be very hot to have a vapor pressure that exceeds atmospheric pressure. Experienced hikers say that water boils at such low temperatures at high altitudes that it does not make for good soup or coffee.
2. The boiling temperature of the salt water is higher than the boiling temperature of the unsalted water because the salt particles lower the water's vapor pressure. More heat is required to boost the lowered vapor pressure to equal the atmospheric pressure.
3. The boiling antifreeze solution boils at a higher temperature than plain water because the particles of antifreeze lower the solution's vapor pressure. This raising of the boiling point is the purpose for using an antifreeze solution in the radiator. It allows the radiator to function more efficiently in cooling the engine.

Freezing Point

When a liquid freezes, its molecules slow to the point that they no longer slide past each other. The molecules of a frozen liquid are arranged in a structure that does not have the freedom of the liquid state. If particles of solute are added to a liquid, the solute particles interfere with the orderly alignment of solvent molecules as freezing takes place. The temperature of the liquid must then be lowered even more before freezing can take place. This phenomenon, called **freezing point depression, means that a liquid with a solute in it freezes at a temperature lower than that at which the liquid alone would freeze.**

EXERCISES

Answer the following questions about freezing point.

1. Why does spreading salt on snow-covered roads cause the snow to melt?
2. A tidal river has salty water for some distance upstream from its mouth. What can you say about the temperature at which this water freezes?
3. As the temperature drops during a winter hiking trip, which freezes first—a canteen of water or a canteen of sweetened tea?

Answers

1. The salt combines with the snow to create a solution whose freezing point is lower than the freezing point of the snow by itself. It is possible for the temperature to become so low that salt does not melt the snow because even ice water freezes when cold enough. Typically salt is ineffective at temperatures lower than 15°F.
2. The salty part of the tidal river, closer to the mouth of the river, freezes at colder temperatures than the less salty river water upstream.
3. The water freezes first because it freezes at a higher temperature than the sweetened tea. The sugar and tea in the water depress its freezing point.

UNIT 8

Atomic Theory and Periodicity

Structure of the Atom; the Periodic Table and Its Relationship to Atomic Structure

The structure of the atom and the organization of the periodic table are so inextricably intertwined as to be inseparable. We will examine the structure of the atom first and then see how this structure determines the arrangement and function of the periodic table. The twentieth century was the setting for the unraveling of many secrets of the atom, with the remainder of its secrets on the scientific horizon. At the beginning of the century, the atom was viewed as a small, solid unit. Early in that century, Ernest Rutherford's experiments showed it to be mostly empty space—a truly startling idea said to be the "shot heard 'round the scientific world." His work led to the conclusion that the atom has a dense, positive center, called the nucleus that is quite small compared to the entire atom. An analogy can be drawn about this size comparison: If the whole atom were the size of Yankee Stadium, the nucleus would be the size of the baseball held by the pitcher.

The experiments of J. J. Thompson revealed the electron to be a negatively charged subatomic particle of almost no mass. Later scientists described the electron's location and behavior within the atom. By 1932 the most elusive of the "big three" subatomic particles, the neutron, was shown to exist by Harold Urey, who received a Nobel Prize for his work. The location of the neutral (no charge) neutron is in the nucleus along with the positively charged proton.

Central to the evolving description of an atom's structure is the electromagnetic spectrum. The word *spectrum* means a range of values. The term electromagnetic describes energy that has both electrical and magnetic properties. Among the electromagnetic energies are the familiar radio, television, visible light, and x-rays. All electromagnetic energy is wavelike, as shown in Figure 8.1.

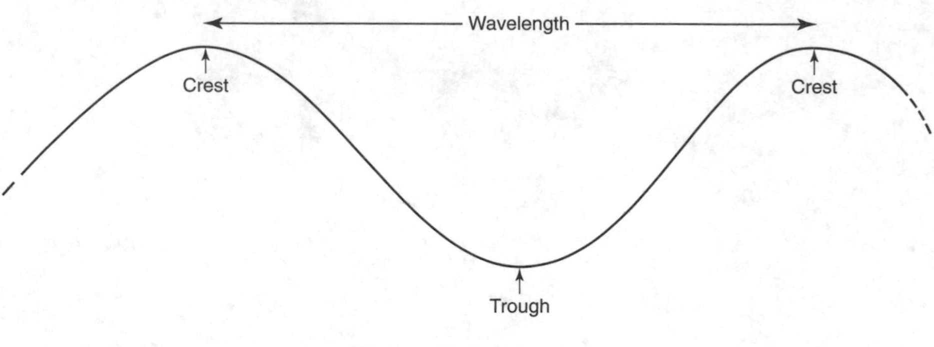

Figure 8.1: Wave

Wavelength is the distance between similar parts of the curve, for example, between crests (tops) or between troughs (bottoms) of adjacent waves. *Frequency* is the number of waves that pass an observation point per second. All types of electromagnetic energy have the same *speed*, which is the speed of light. The diagram of the electromagnetic spectrum in Figure 8.2 is arranged in order of decreasing wavelength and therefore increasing frequency and increasing energy.

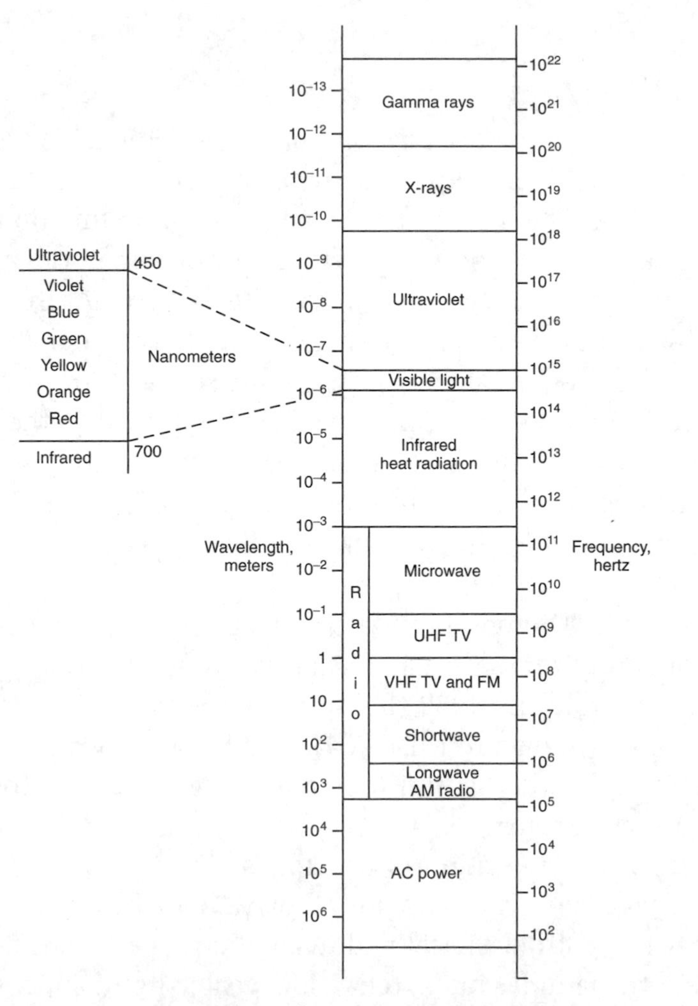

Figure 8.2: Electromagnetic spectrum

EXERCISES

Examine the electromagnetic spectrum in Figure 8.2 in order to answer these questions.

1. Which has the longer wavelength, radio waves or microwaves?
2. Which has more energy, x-rays or cosmic rays?
3. Are sound waves electromagnetic in nature?
4. Is there a difference in the speeds of radio and light?

Answers

1. Radio waves have a longer wavelength.
2. Cosmic rays have more energy.
3. Sound waves are not part of the electromagnetic spectrum.
4. No; all electromagnetic waves travel at the speed of light.

VISIBLE LIGHT

Visible light is the part of the electromagnetic spectrum that is vitally important in understanding the atom. The following memory device helps us remember the order of the colors within the visible light part of the spectrum: ROY G. BIV, which stands for red, orange, yellow, green, blue, indigo, violet. This array of colors is caused by slight variations in the wavelengths, with red light having the longest wavelength of visible light and violet having the shortest. Red light also has the least amount of energy of the visible wavelengths. Many years ago there was a public outcry when police vehicles began to use the now familiar blue lights. There were detractors who saw blue police lights as not being sufficiently strong and powerful, not knowing that blue light has more energy. If you wonder why violet would not be an even better choice, the answer lies in the relative weakness of the human eye's response to violet light (Figure 8.3).

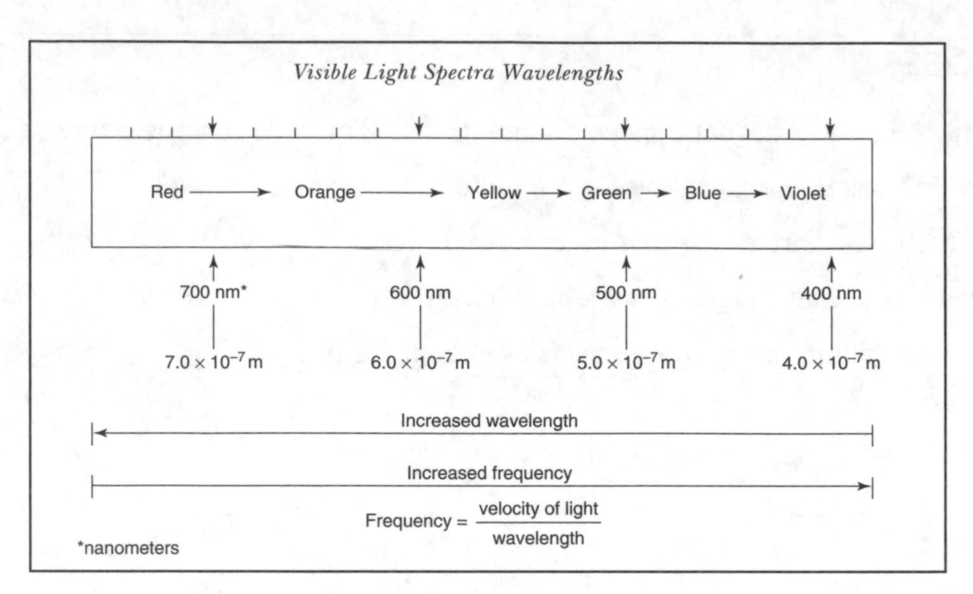

Figure 8.3: Visible-light portion of the spectrum

As you think about the colors of visible light, imagine crayons in a very large crayon box. There are many crayons, for example, that you could call red. In the electromagnetic spectrum there are many wavelengths of light that could be called red even though they are slightly different from one another.

EXERCISES

Answer the following questions about visible light.

1. Of green and yellow light, which has the longer wavelength?
2. Of violet and blue light, which has more energy?
3. Which of the visible colors has the highest frequency?

Answers

1. yellow
2. violet
3. violet

LIGHT AND THE STRUCTURE OF THE ATOM

From the very ancient art of fireworks come clues about the location and behavior of electrons in atoms. The different colors of fireworks are the result of the actions of electrons within their compounds when ignited. Before fireworks are set off, the electrons are in what is called the *ground state,* which is their normal location within the atom. Once the fireworks are ignited, some of the electrons acquire enough energy to go to the *excited state*. Much like a jumper on a trampoline, electrons cannot stay at this elevated position, and so they fall back to a lower position. This fall results in their giving up energy. Remembering that each color of visible light is associated with a particular energy, you could say that each color is caused by a different energy given up when electrons fall back to a lower energy from an excited state (Figure 8.4).

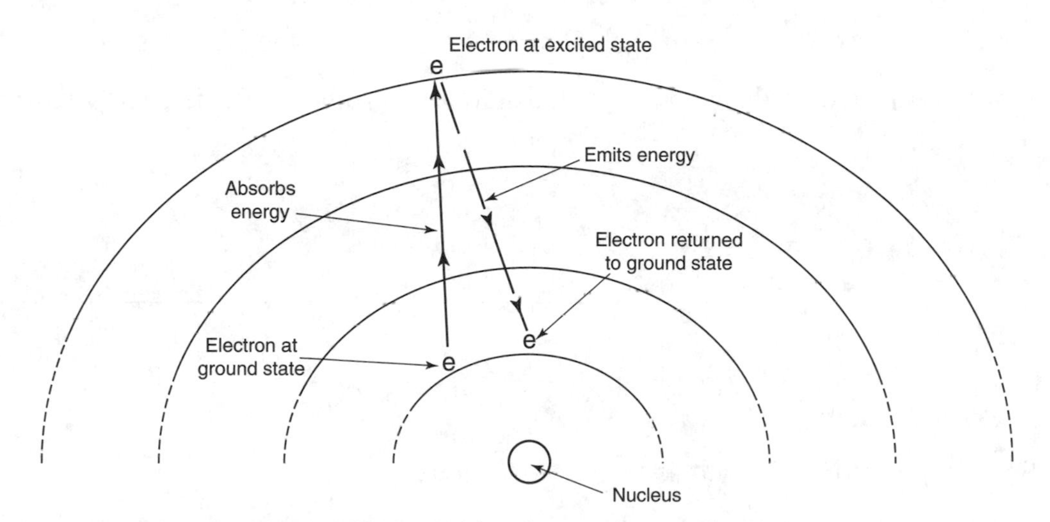

Figure 8.4: Excited and ground-state electrons

The fact that excited atoms give off specific colors and not a rainbow of colors suggested to Niels Bohr, a Danish physicist, that electrons are permitted in only certain locations within the atom. These locations are called *energy levels*. Each element behaves in its own unique way when excited by heat or electricity and produces a very specific pattern of lines of color called the *atomic spectrum* of that element (Figure 8.5). This unique chemical fingerprint is the foundation of atomic spectroscopy, a method of analysis used by forensic and medical laboratories to identify elements

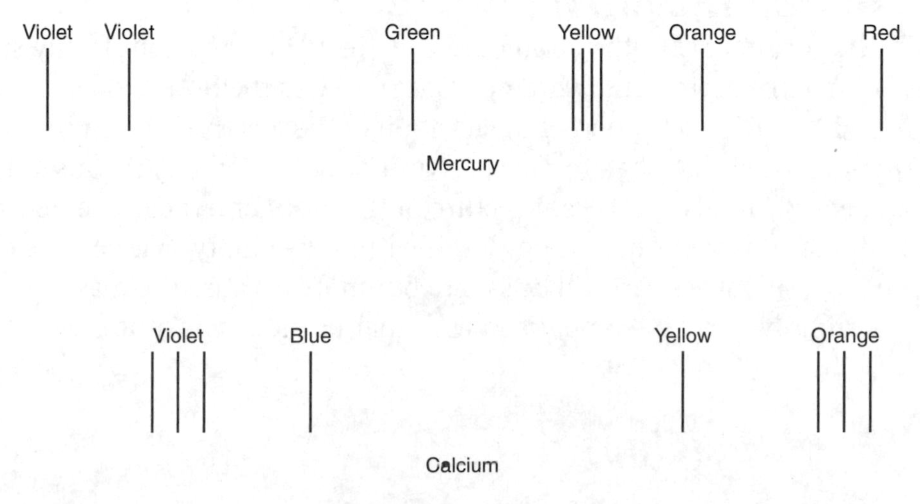

Figure 8.5: Atomic spectra

EXERCISES

Answer these questions about electrons.

1. Are electrons in auto tires in the ground state or in an excited state?
2. What makes electrons "leap" from a ground state to an excited state?
3. If every atomic spectrum were a rainbow instead of a unique pattern of light, what could be said about the locations of electrons in the atom?
4. What is at least one cause of each element having a unique atomic spectrum?
5. When an electron falls from an excited state to a ground state, is energy required or released?

Answers

1. The ground state
2. Some type of energy, such as heat or electricity.
3. That electrons are not confined to energy levels but can be randomly located throughout the atom.
4. Each element has a different number of electrons, a different-sized nucleus to pull the electrons back to the ground state, and different distances from the nucleus to the outer electrons.
5. Energy is released.

BOHR MODEL OF THE ATOM

Just as the Rutherford model of the atom developed in 1911 was scientifically startling with its revelation of the atom as mostly empty space, so was the Bohr model of the atom introduced in 1913 with its definition of the location of the electron within the atom. As Bohr and others realized that the atomic spectrum of each element is caused by electrons changing energy levels, a different picture of the atom emerged. The new picture of the atom had electrons at various energy levels within the empty space of Rutherford's model (Figure 8.6). This space can still be said to be empty because the mass of the electrons is extraordinarily small in comparison with that of the whole atom.

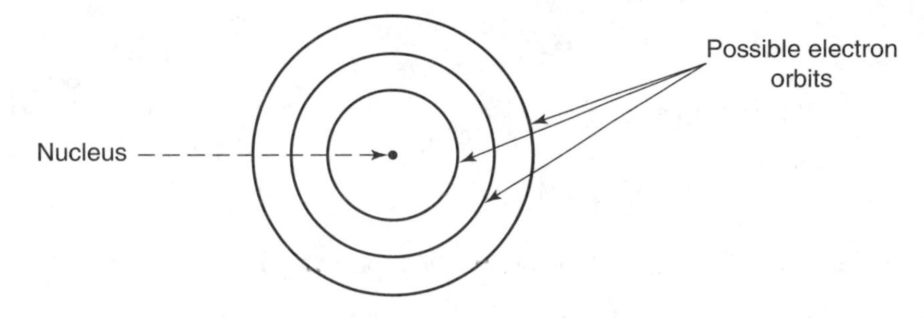

Figure 8.6: Bohr model

The Bohr model was a huge breakthrough, departing from the accepted thinking of the time, which was that electrons were randomly scattered throughout the atom. Bohr realized that the evidence provided by atomic spectra, as in fireworks, meant that electrons were restricted to being in energy levels. Excellent though this model was for very simple atoms, it was not adequate for larger, more complex atoms. As scientific thought and experimentation evolved, a new kind of physics called quantum mechanics was developed.

Quantum mechanics is a highly mathematical view of the atom and expands the classical physics viewpoint to explain atomic structure. A staircase is a useful analogy in discussing quanta, in that you climb the stairs in certain quanta or in certain discrete units, namely, the steps themselves. You cannot step anywhere other than on a stair tread, and standing in between steps is not possible. In the same way, electrons have certain permitted locations and cannot exist between these locations.

QUANTUM MECHANICAL MODEL OF THE ATOM

As scientists worked on the creation of atomic models that better explained experimental data, it became apparent that mathematical equations and statistical probability were more useful than actual physical models. Instead of thinking of an electron as a particle traveling in a prescribed path around the nucleus of an atom, scientists began to consider the probability of the electron's location. This is analogous to taking pictures of a basketball game in progress. Most of the pictures show the ball in the vicinity of either basket, so these are the areas of high probability. At any one instant, we cannot be certain of the location of the ball, but we know the probability of its likely locations. Similarly, an electron does not travel in a prescribed path but moves randomly within an area of probability. This randomness can be likened to the flitting of a butterfly in a flower garden. The butterfly does not have a definite path, but it is in constant motion in the garden.

A German physicist, Werner Heisenberg, formalized this idea with his *uncertainly principle*, which says that we cannot know the exact location or motion of an electron. Although we cannot describe the path of an electron's motion, we can speak of the probability of finding it in a given location at any time. In Figure 8.7, the darker areas represent a greater probability that the electron is present in this location.

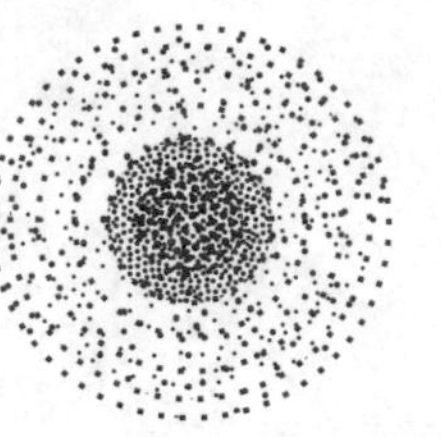

Figure 8.7: The probability of an electron being in a given location.

EXERCISES

Answer these questions dealing with the model of the atom.

1. How does the idea of probability differ from the idea behind Bohr's model of electron behavior?
2. How are the Bohr model and the quantum mechanical model of the atom similar?

Answers

1. The electron in Bohr's model has a definite path of travel, much like that of a race car on a track, whereas the electron in the probability model moves in a random manner and the best description of its location is a statistical likelihood.
2. The similarity between these models is the idea of energy levels and the location of the electron.

Quantum Numbers

Each electron can be assigned a set of four quantum numbers. In a sense, these numbers are the electron's address. An explanation of each quantum number follows.

The First Quantum Number

The first quantum number is called the *principal quantum number* and describes the energy level. These quantum numbers are integers beginning with 1. The first energy level is closest to the nucleus, and each successive energy level is farther from the nucleus (Figure 8.8).

$n = 5$ - - - - - - - - - - - - - - - - -

$n = 4$ - - - - - - - - - - - - - - - - - -

$n = 3$ - - - - - - - - - - - - - - - - - - -

$n = 2$ -

$n = 1$ -

Figure 8.8: Energy levels

The Second Quantum Number

The second quantum number describes the shape of the orbital as *s*, *p*, *d*, *f*, or *g*. These shapes do not describe the electron's path but rather are mathematical models showing the probability of the electron's location. The *s* and *p* orbital shapes are shown in Figure 8.9, but descriptions of the *d* and *f* orbitals are reserved for more advanced texts.

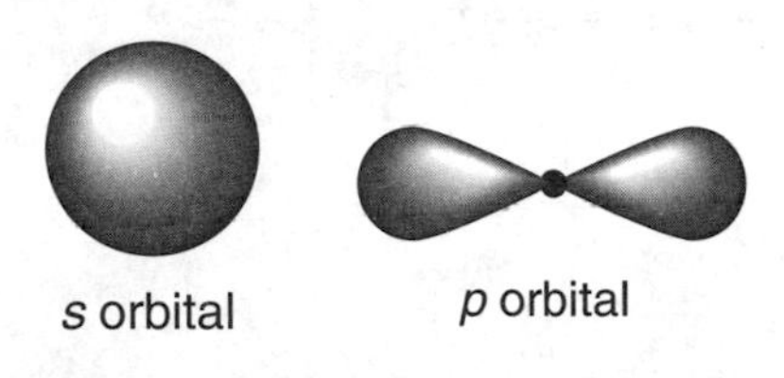

Figure 8.9: Orbital shapes

The *s*, *p*, *d*, and *f* levels are also called energy sublevels. The *s* sublevel has one orbital, an *s* orbital. The *p* sublevel has three orbitals, *x*, *y*, and *z*. The *d* sublevel has five orbitals and therefore a total capacity of ten electrons. Examine Figure 8.10 to see how the sublevel idea fits in with the concept of energy levels.

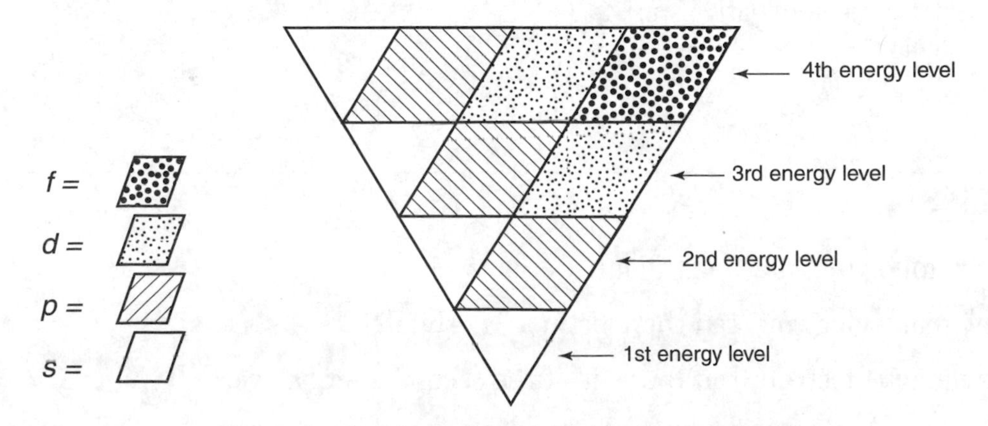

Figure 8.10: Principal levels divided into sublevels

Notice in Table 8.1 that each energy level farther from the nucleus contains one more sublevel than the one before it. Each sublevel has a different electron capacity.

TABLE 8.1: ENERGY SUBLEVELS

s Sublevel	*p* Sublevel	*d* Sublevel
2-electron capacity	6-electron capacity	10-electron capacity
1 orbital	3 orbitals	5 orbitals
At every energy level	at every energy level after the first level	at every energy level after the second level

The Third Quantum Number

The third quantum number describes the orientation in space of an orbital. For example, in describing a *p* orbital, this third quantum number (*x*, *y*, or *z*) gives its orientation in space. To help visualize this idea, think of the *x* and *y* axes on graph paper. The *z* axis is at a right angle to the plane of the paper, as if you have impaled the graph paper on a pencil (Figure 8.11).

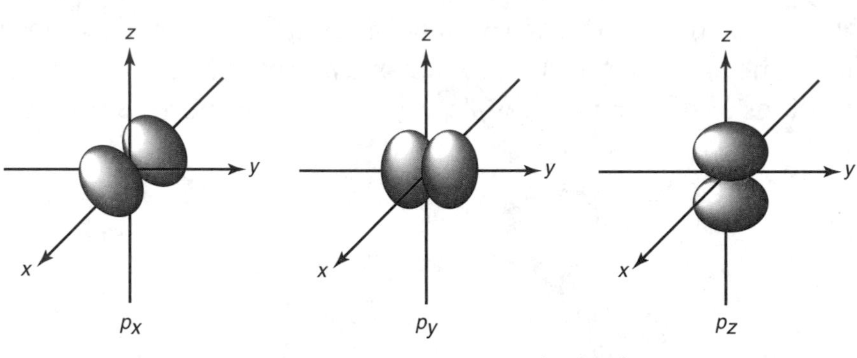

Figure 8.11: Orientation of orbitals

The Fourth Quantum Number

The fourth quantum number is the spin number. Electrons are said to have either a +½ spin or a –½ spin, which we might think of as clockwise and counterclockwise, respectively. Within any orbital, the first electron is said to have a positive spin, whereas the second electron has a negative spin.

EXERCISES

Answer the following questions about energy levels.

1. How many electrons can there be in a 3*s* orbital?
2. How many electrons can there be at the second energy level?

3. How many sublevels are there at the second energy level?
4. How many electrons can there be in any *p* orbital?
5. Is the 4*s* orbital the same size as the 3*s* orbital?
6. What are the three *p* orbitals called?

Answers

1. Two. Every *s* orbital has a maximum capacity of two electrons.
2. Eight. Two in the *s* orbital and six in the *p* orbital.
3. Two. The *s* orbital and the *p* orbital.
4. Six. Every *p* orbital has a capacity of six electrons.
5. No. The farther the energy levels are from the nucleus, the larger they are, in much the same way that rings on a target are larger the farther they are from the bull's-eye.
6. The three *p* orbitals are the *x*, *y*, and *z*, each capable of holding two electrons.

The extraordinary minds of Albert Einstein, Neils Bohr, Marie Curie, Enrico Fermi, and countless others provided the insight, experimentation, and sheer brain power needed to illuminate the nature of the atom. Their legacy provides a spectacular foundation for the future unraveling of atomic mysteries.

PERIODICITY

Initially, the known elements were organized in an arrangement based on similarities in properties (reactions, appearance, melting points, and so on). Dimitri Mendeleev, a Russian scientist, is credited with the most complete of the early arrangements of what is now called the periodic table. In 1869 scientists did not know about energy levels or *s*, *p*, *d*, and *f* sublevels. As you work your way through the next pages, you will see how these ideas about energy levels and sublevels not only fit in with the periodic table but are also responsible for its arrangement.

Electron Configuration

Writing what is called the *electron configuration* is a way of describing each of an element's electrons. It requires an *Aufbau diagram* (Figure 8.12) and knowledge of the number of electrons in a particular atom. The number of electrons is shown on the periodic table as the smaller of the two numbers in the square for each element.

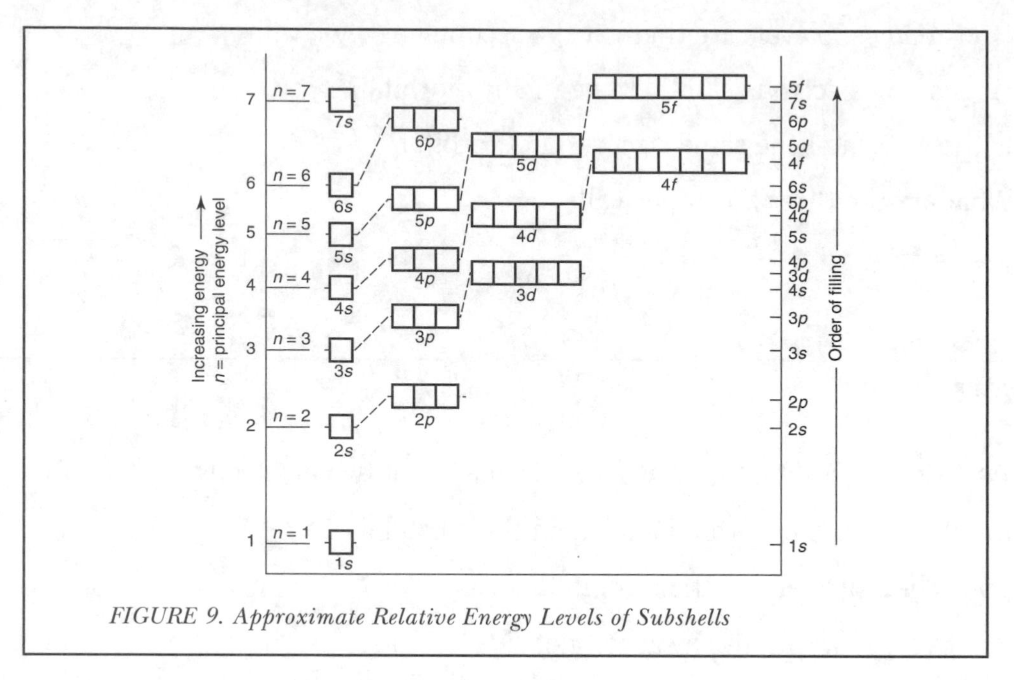

FIGURE 9. Approximate Relative Energy Levels of Subshells

Figure 8.12: Aufbau diagram

An electron configuration can be written by following these steps:

- Find out the number of electrons for the element.
- Start with the 1*s* part of the Aufbau diagram.
- For each orbital, insert the maximum number of electrons in the exponent position.
- Continue until each of the element's electrons has been described.

The electron configuration for the element sodium, which has 11 electrons, looks like this:

$$1s^2 2s^2 2p^6 3s^1$$

This electron configuration provides a variety of information:

- The sum of the superscript numbers (2 + 2 + 6 + 1 = 11) gives the number of electrons in an atom of sodium.
- The coefficients show that the sodium atom has three principal energy levels because the largest of these coefficients is a 3.
- There is one electron in the outer energy level ($3s^1$).

EXERCISES

Answer the following questions about electron configuration.

1. What is the electron configuration for potassium (K).
2. What is the electron configuration for rubidium (Rb).
3. What two features do sodium, potassium, and rubidium have in common?
4. What generalized statement can you make about the elements in the first column of the periodic table?
5. The next-to-last column of elements in the periodic table begins with fluorine. Write the electron configuration for fluorine (F).
6. Now, write the electron configuration for chlorine (Cl).
7. What general statement can you make about elements in this column of the periodic table?

Answers

1. $1s^2 2s^2 2p^6 3s^2 3p^6 4s^1$
2. $1s^2 2s^2 2p^6 3s^2 3p^6 4s^2 3d^{10} 4p^6 5s^1$
3. They are in column 1 of the periodic table, and they all end with s^1.
4. In general, you could say that all the elements in column 1 of the periodic table have one electron in the outermost energy levels and that the number of energy level increases by 1 for each element from top to bottom.
5. $1s^2 2s^2 2p^5$
6. $1s^2 2s^2 2p^6 3s^2 3p^5$
7. The outermost electrons of all elements in this column of the periodic table have the pattern s^2p^5.

PERIODICITY AND ELECTRON CONFIGURATION

If you write the electron configuration for every element in the periodic table, you will arrive at the conclusions shown in Figure 8.13.

Periodic Table of the Elements

s block =

p block =

d block =

f block =

1A	2A	3	4	5	6	7	8	9	10	11	12	3A	4A	5A	6A	7A	Noble Gases
										1 H 1.008							2 He 4.00
3 Li 6.94	4 Be 9.01											5 B 10.81	6 C 12.01	7 N 14.01	8 O 16.00	9 F 19.00	10 Ne 20.18
11 Na 22.99	12 Mg 24.31											13 Al 26.98	14 Si 28.09	15 P 30.97	16 S 32.06	17 Cl 35.45	18 Ar 39.95
19 K 39.10	20 Ca 40.08	21 Sc 44.96	22 Ti 47.90	23 V 50.94	24 Cr 52.00	25 Mn 54.94	26 Fe 55.85	27 Co 58.93	28 Ni 58.71	29 Cu 63.54	30 Zn 65.37	31 Ga 69.72	32 Ge 72.59	33 As 74.92	34 Se 78.96	35 Br 79.91	36 Kr 83.80
37 Rb 85.47	38 Sr 87.62	39 Y 88.91	40 Zr 91.22	41 Nb 92.91	42 Mo 95.94	43 Tc 98.91	44 Ru 101.07	45 Rh 102.91	46 Pd 106.04	47 Ag 107.87	48 Cd 112.40	49 In 114.82	50 Sn 118.69	51 Sb 121.75	52 Te 127.60	53 I 126.90	54 Xe 131.30
55 Cs 132.91	56 Ba 137.34	57 La 138.91	72 Hf 178.49	73 Ta 180.95	74 W 183.85	75 Re 186.2	76 Os 190.2	77 Ir 192.2	78 Pt 195.09	79 Au 196.97	80 Hg 200.59	81 Tl 204.37	82 Pb 207.19	83 Bi 208.98	84 Po 209	85 At 210	86 Rn 222
87 Fr 223	88 Ra 226.03	89 Ac 227.03	104 Unq	105 Unp	106 Unh	107 Uns	108 Uno	109 Une									

Lathanides	58 Ce 140.12	59 Pr 140.91	60 Nd 144.24	61 Pm 146.92	62 Sm 150.35	63 Eu 151.96	64 Gd 157.25	65 Tb 158.92	66 Dy 162.50	67 Ho 164.93	68 Er 167.26	69 Tm 168.92	70 Yb 173.04	71 Lu 174.97
Actinides	90 Th 232.04	91 Pa 231.04	92 U 238.03	93 Np 237.05	94 Pu 150.35	95 Am 241.06	96 Cm 247.07	97 Bk 249.08	98 Cf 251.09	99 Es 254.09	100 Fm 257.10	101 Md 258.10	102 No 255	103 Lr 257

Figure 8.13: Periodic table showing *s, p, d,* and *f* blocks.

VALENCE ELECTRONS

Valence electrons are the outermost electrons in an atom. For example, sodium has the electron configuration $1s^22s^22p^63s^1$, and so its outermost energy level is level 3. In energy level 3 there is one electron, and so sodium has one valence electron, the $3s^1$. As you review the electron configurations you wrote for the other elements in column 1, notice that they all have one valence electron.

EXERCISES

In each of the following electron configurations circle the valence electrons, remembering that they are in the highest principal energy level. Then state the number of valence electrons.

1. $1s^22s^22p^63s^23p^4$. Number of valence electrons?
2. $1s^22s^22p^63s^23p^64s^23d^{10}4p^3$. Number of valence electrons?
3. $1s^22s^22p^63s^23p^64s^23d^{10}4p^6$. Number of valence electrons?

Answers

1. $1s^22s^22p^63s^23p^4$ 6 (2 in the 3*s* plus 4 in the 3*p*)
2. $1s^22s^22p^63s^23p^64s^23d^{10}4p^3$ 5 (2 in the 4*s* plus 3 in the 4*p*)
3. $1s^22s^22p^63s^23p^64s^23d^{10}4p^6$ 8 (2 in the 4*s* plus 6 in the 4*p*)

METALS AND NONMETALS

It is the valence electrons that influence the way an element reacts. Metals are elements that lose electrons, whereas nonmetals are elements that gain electrons. The valence electron pattern of elements in the last column in the periodic table is s^2p^6. This extremely stable electron configuration gives rise to the group name for these elements. They are called *noble gases* because they do not react with other elements except under extraordinary circumstances. It is this stability that other elements "want," and so they gain or lose electrons in order to achieve the s^2p^6 configuration. The *octet rule* refers to this noble gas configuration because there are eight electrons in the outer energy level.

Nonmetals

The electron configuration for chlorine is $1s^22s^22p^63s^23p^5$. In order to achieve the configuration of a noble gas, chlorine must gain one electron and become $1s^22s^22p^63s^23p^6$. Recall from your studies of the formation of ions that this gain of one electron produces a –1 ion: Cl–.

In general, nonmetals are elements that are to the right of the stairstep line in the periodic table and need to gain electrons in order to achieve an electron configuration like that of a noble gas (Figure 8.14).

Periodic Table of the Elements

1A	2A	3	4	5	6	7	8	9	10	11	12	3A	4A	5A	6A	7A	Noble Gases
										1 H 1.008							2 He 4.00
3 Li 6.94	4 Be 9.01											5 B 10.81	6 C 12.01	7 N 14.01	8 O 16.00	9 F 19.00	10 Ne 20.18
11 Na 22.99	12 Mg 24.31											13 Al 26.98	14 Si 28.09	15 P 30.97	16 S 32.06	17 Cl 35.45	18 Ar 39.95
19 K 39.10	20 Ca 40.08	21 Sc 44.96	22 Ti 47.90	23 V 50.94	24 Cr 52.00	25 Mn 54.94	26 Fe 55.85	27 Co 58.93	28 Ni 58.71	29 Cu 63.54	30 Zn 65.37	31 Ga 69.72	32 Ge 72.59	33 As 74.92	34 Se 78.96	35 Br 79.91	36 Kr 83.80
37 Rb 85.47	38 Sr 87.62	39 Y 88.91	40 Zr 91.22	41 Nb 92.91	42 Mo 95.94	43 Tc 98.91	44 Ru 101.07	45 Rh 102.91	46 Pd 106.04	47 Ag 107.87	48 Cd 112.40	49 In 114.82	50 Sn 118.69	51 Sb 121.75	52 Te 127.60	53 I 126.90	54 Xe 131.30
55 Cs 132.91	56 Ba 137.34	57 La 138.91	72 Hf 178.49	73 Ta 180.95	74 W 183.85	75 Re 186.2	76 Os 190.2	77 Ir 192.2	78 Pt 195.09	79 Au 196.97	80 Hg 200.59	81 Tl 204.37	82 Pb 207.19	83 Bi 208.98	84 Po 209	85 At 210	86 Rn 222
87 Fr 223	88 Ra 226.03	89 Ac 227.03	104 Unq	105 Unp	106 Unh	107 Uns	108 Uno	109 Une									

Lathanides	58 Ce 140.12	59 Pr 140.91	60 Nd 144.24	61 Pm 146.92	62 Sm 150.35	63 Eu 151.96	64 Gd 157.25	65 Tb 158.92	66 Dy 162.50	67 Ho 164.93	68 Er 167.26	69 Tm 168.92	70 Yb 173.04	71 Lu 174.97
Actinides	90 Th 232.04	91 Pa 231.04	92 U 238.03	93 Np 237.05	94 Pu 150.35	95 Am 241.06	96 Cm 247.07	97 Bk 249.08	98 Cf 251.09	99 Es 254.09	100 Fm 257.10	101 Md 258.10	102 No 255	103 Lr 257

Figure 8.14

Oxygen is in a column in the nonmetal category, and it has eight electrons. Its electron configuration is $1s^22s^22p^4$. The nearest noble gas configuration is $1s^22s^22p^6$; so oxygen gains two electrons, becoming a –2 ion.

Metals

The electron configuration for magnesium is $1s^22s^22p^63s^2$. The desirable configuration is s^2p^6, but the coefficient in front of each must be the same. If the magnesium atom loses two electrons, the stable configuration of $1s^22s^22p^6$ results. A loss of electrons is the hallmark of a metal. The number of electrons that it loses is determined by what it must do in order to resemble a noble metal in its electron configuration.

EXERCISES

Examine each of the following electron configurations and then decide how many electrons must be lost or gained to reach a stable electron configuration. Is the element in question a metal or a nonmetal?

Electron Configuration	No. of Electrons Lost or Gained	Metal or Nonmetal?
1. $1s^22s^22p^3$		
2. $1s^22s^22p^63s^23p^64s^1$		
3. $1s^22s^22p^63s^23p^64s^23d^{10}4p^1$		
4. $1s^22s^22p^63s^23p^64s^23d^{10}4p^4$		

Answers

1. Three electrons are gained to become $1s^22s^22p^6$, making it a nonmetal.
2. One electron is lost to become $1s^22s^22p^63s^23p^6$, making it a metal.
3. Three electrons are lost to become $1s^22s^22p^63s^23p^63d^{10}$, making it a metal.
4. Two electrons are gained to become $1s^22s^22p^63s^23p^64s^23d^{10}4p^6$, making it a nonmetal.

As expected, the elements of a given column in the periodic table all behave in the same way because they all have the same number of valence electrons. There are a few exceptions to this rule, as you can tell if you look at the stairstep line passing through the periodic table. The elements to the right of the stairstep behave as nonmetals, whereas the elements to the left of this line behave as metals. But overall, the generalizations and trends apparent from the periodic table make it invaluable as a predictive tool.

Chemical Bonding

Ionic and Covalent Bonding; Lewis Structures; Bond Character; Single, Double, and Triple Bonds

Chemical bonds are the forces that hold atoms or ions together to make all molecules and compounds. The nature of these bonds determines the properties of the substances that are created. Chemical bonds are a function of the valence electrons of the atoms involved in the bond.

ELECTRONEGATIVITY

Electronegativity is a measure of the ability of an atom to attract bonding electrons. The strength of this attraction is based on two factors: the distance from the nucleus to the outermost electrons, and the valence electron pattern. The distance factor is a function of the fact that the closer the nucleus is to the outer electrons, the greater the power of the nucleus in pulling in other electrons. This is similar to the observation that a magnet works best when close to an object. As the valence electron pattern approaches the noble gas valence electron pattern, the more effective the element is at attracting electrons. Noble gases have virtually no electronegativity, as they rarely react.

To see how these two factors influence electronegativity, turn to the periodic table and locate the elements fluorine (F) and francium (Fr). Fluorine has the most electronegativity of any element. It is very close to having the valence electron pattern of a noble gas with its $1s^22s^22p^5$, the nearest noble gas being $1s^22s^22p^6$. In addition, fluorine has the smallest atoms of the elements in its column because, at two, it has the fewest number of energy levels. Considering both these factors, it makes sense that fluorine is the most electronegative of the elements.

Now look at francium (Fr), located at the bottom of the first column. In terms of valence electron pattern, francium has an s^1, meaning that it would have to gain seven electrons to become s^2p^6. Far more efficient for francium would be to lose one electron, thereby achieving the s^2p^6 of its next lower energy level. Because francium "prefers" to lose electrons rather than gain electrons, its electron attracting power is very low. As far as the distance factor goes, francium is the largest atom in its column, and so its nucleus is far from the outermost electrons. This distance also diminishes the ability of its nucleus to attract new electrons.

EXERCISES

Answer the following questions dealing with electronegativity.

1. Using the periodic table, arrange these three elements in order of increasing electronegativity: S, Cl, Se.
2. Which would you expect to have higher electronegativity: sodium or potassium?
3. In the case of sodium and potassium, which factor is responsible for your answer to question 2?
4. In general, would you expect metals or nonmetals to have higher electronegativity?

Answers

1. Se, S, Cl. Se has the lowest electronegativity because it has the largest atomic size, meaning that its nucleus is farther from the outer electrons. Chlorine is closer to having the noble gas valence electron pattern than is sulfur.
2. Sodium has a higher electronegativity than potassium.
3. Because sodium and potassium both have the same valence electron pattern, with s^1, it must be the distance factor that is important.
4. Nonmetals have higher electronegativity because they are gainers of electrons, this being their best pathway to achieving a noble gas valence electron pattern.

Using Electronegativity

Use of the idea of electronegativity would certainly be facilitated by a periodic table with all of the electronegativity values listed. Use Figure 9.1 as you work with bonding issues.

H 2.1																
Li 1.0	Be 1.5											B 2.0	C 2.5	N 3.1	O 3.5	F 4.0
Na 1.0	Mg 1.3											Al 1.5	Si 1.8	P 2.1	S 2.4	Cl 2.9
K 0.8	Ca 1.1	Sc 1.2	Ti 1.3	V 1.5	Cr 1.6	Mn 1.6	Fe 1.7	Co 1.7	Ni 1.8	Cu 1.8	Zn 1.7	Ga 1.8	Ge 2.0	As 2.2	Se 2.5	Br 2.8
Rb 0.8	Sr 1.0	Y 1.1	Zr 1.2	Nb 1.3	Mo 1.3	Tc 1.4	Ru 1.4	Rh 1.5	Pd 1.4	Ag 1.4	Cd 1.5	In 1.5	Sn 1.7	Sb 1.8	Te 2.0	I 2.5
Cs 0.7	Ba 0.9	La 1.1	Hf 1.2	Ta 1.4	W 1.4	Re 1.5	Os 1.5	Ir 1.6	Pt 1.5	Au 1.4	Hg 1.5	Tl 1.5	Pb 1.6	Bi 1.7	Po 1.8	At 2.2
Fr 0.7	Ra 0.9	Ac 1.0														

Figure 9.1 Electronegativity values

Bond Character

The difference between the electronegativity values of the two atoms involved in creating a chemical bond explains the ideas of bond character, bond polarity, covalent bonds, and ionic bonds. Figure 9.2 illustrates these concepts.

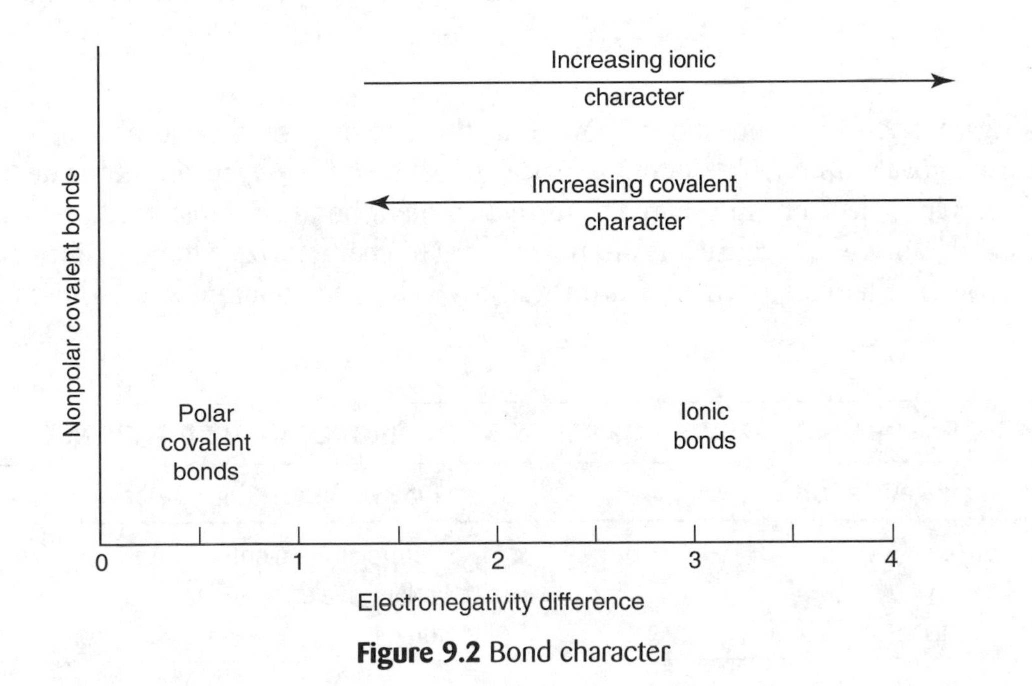

Figure 9.2 Bond character

EXERCISES

Examine Figure 9.2 in order to answer the following questions.

1. As the electronegativity difference between bonding atoms increases, what happens to the ionic character of their bond?
2. What kind of bond results from an electronegativity difference of zero?
3. Can a nonpolar bond be an ionic bond?
4. If the electronegativity difference between two bonding atoms is 1.0, can the bond be characterized as having a more ionic or a more covalent character?

Answers

1. The ionic character increases with increasing electronegativity difference.
2. An electronegativity difference of zero results in a nonpolar bond.
3. A nonpolar bond is not an ionic bond. It is a covalent bond.
4. An electronegativity difference of 1.0 indicates more covalent character than ionic character.

Working through these questions shows that there is no arbitrary number separating ionic from covalent bonding. Rather we speak of bond *character*, meaning the degree to which bonding electrons are shared to form a covalent bond or transferred to create an ionic bond. However, because it is often necessary to characterize a bond as either covalent or ionic, chemists have agreed on number values for making such a distinction (Table 9.1).

TABLE 9.1: CHARACTERIZATION OF BONDS BY ELECTRONEGATIVITY DIFFERENCE

Electronegativity Difference	Type of Bond
0.0 to 0.2	nonpolar covalent
0.3 to 1.4	polar covalent
Greater than 1.5	ionic

When using these values to characterize a bond, do not consider the subscripts in the chemical formula. That is, for H_2O, subtract the electronegativity of one H (2.1) from that of one O (3.5) to get a difference of 1.4. The bonds in water are therefore said to be polar covalent bonds. The smaller value should always be subtracted from the larger so that the difference is always positive.

EXERCISE

For each of the following, determine if the bond(s) is(are) nonpolar covalent, polar covalent, or ionic.

1. SO_2
2. CCl_4
3. Cl_2
4. $CaCl_2$
5. O_3
6. MgS
7. AlF_3

Answers

1. 3.5 – 2.5 = 1.0; polar covalent
2. 3.0 – 2.5 = 0.5; polar covalent
3. 3.0 – 3.0 = 0; nonpolar covalent
4. 3.0 – 1.0 = 2.0; ionic
5. 3.5 – 3.5 = 0; nonpolar covalent
6. 2.5 – 1.2 = 1.3; polar covalent
7. 4.0 – 1.5 = 2.5; ionic

LEWIS SYMBOLS

Writing electron configurations to illustrate what happens to the electrons in bonding is an unwieldy process. Because it is the valence electrons that participate in an atom's reactions, a symbol can be used that includes only these electrons. G. N. Lewis, a noted chemist in the early 1900s, devised such a system. In order to write the *Lewis symbol* for any element, follow these steps.

- Write the chemical symbol for the element.
- Find its number of valence electrons. This number is the same as the column number for the group A elements in the periodic table.
- Place as many bold dots around the symbol as there are valence electrons.
- It is a convention that the first four dots are placed one on each of the four sides of the symbol, with any remaining dots paired up with the first dots.

Note the following examples of the Lewis symbols.

Periodic table column:	1A	2A	3A	4A	5A	6A	7A	8A
Lewis symbols:	•Na	•Mg•	•Al• (one dot above)	•Si• (one dot above, one below)	•P• (two dots above, one below)	•S: (two dots above, one below)	:Cl• (two dots above, one below)	:Ar: (two dots above, two below)

EXERCISES

Write the Lewis symbol for each of the following elements.

1. calcium
2. nitrogen
3. bromine
4. potassium
5. boron
6. oxygen
7. carbon
8. neon

Answers

1. •Ca•
2. •N: (one dot above, one below)
3. :Br• (two dots above, two below)
4. K•
5. •B• (one dot above)
6. :O• (two dots above, one below)
7. •C• (one dot above, one below)
8. :Ne: (two dots above, two below)

IONIC BONDS

Ionic bonds result from the interaction of two or more atoms whose electronegativity difference is greater than 1.5. Ionic bonds are characterized by a gain of electrons by the more electronegative atom and a loss of electrons by the less electronegative atom. The result of this gain and loss is the formation of positive and negative ions.

Using Lewis Symbols for Ionic Bonds

In ionic bonding, atoms gain and lose electrons in order to achieve a noble gas electron configuration. Here's how Lewis symbols show this electron gain and loss in the ionic bonding of sodium chloride:

$Na\bullet \longrightarrow \bullet\ddot{\underset{\bullet\bullet}{Cl}}\!:$ $\qquad$ Na^{+} $\qquad$ $:\ddot{\underset{\bullet\bullet}{Cl}}^{-}\!:$

Observe that the sodium loses one electron and the chlorine gains one electron. This one-to-one ratio is not always the case. Suppose that the reaction under consideration is between magnesium and chlorine. Writing the Lewis symbols for each shows that magnesium has two valence electrons to give away, whereas chlorine has a vacancy for only one. Therefore, another chlorine atom is needed to provide a place for the second available magnesium electron. This is described in Lewis symbols as

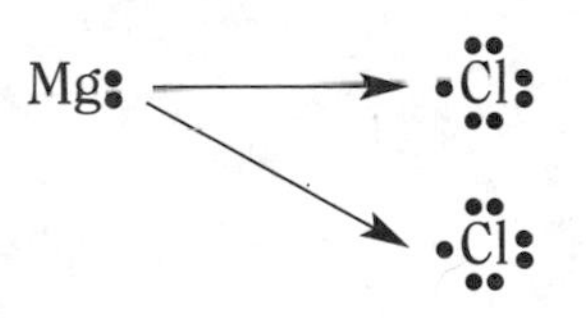

The resulting chemical formula for magnesium chloride is $MgCl_2$. Remember that the number of electrons given up must equal the number of electrons acquired.

EXERCISES

Use Lewis symbols to show the transfer of electrons in each of the following combinations. This will allow you to see how many atoms of each element must be present so that you can then write the correct formula for the compound.

1. calcium and fluorine
formula:

2. potassium and chlorine
formula:

3. sodium and oxygen
formula:

4. aluminum and bromine
formula:

5. magnesium and sulfur
formula:

6. aluminum and oxygen
formula:

7. barium and nitrogen
formula:

8. silicon and oxygen
formula:

Answers

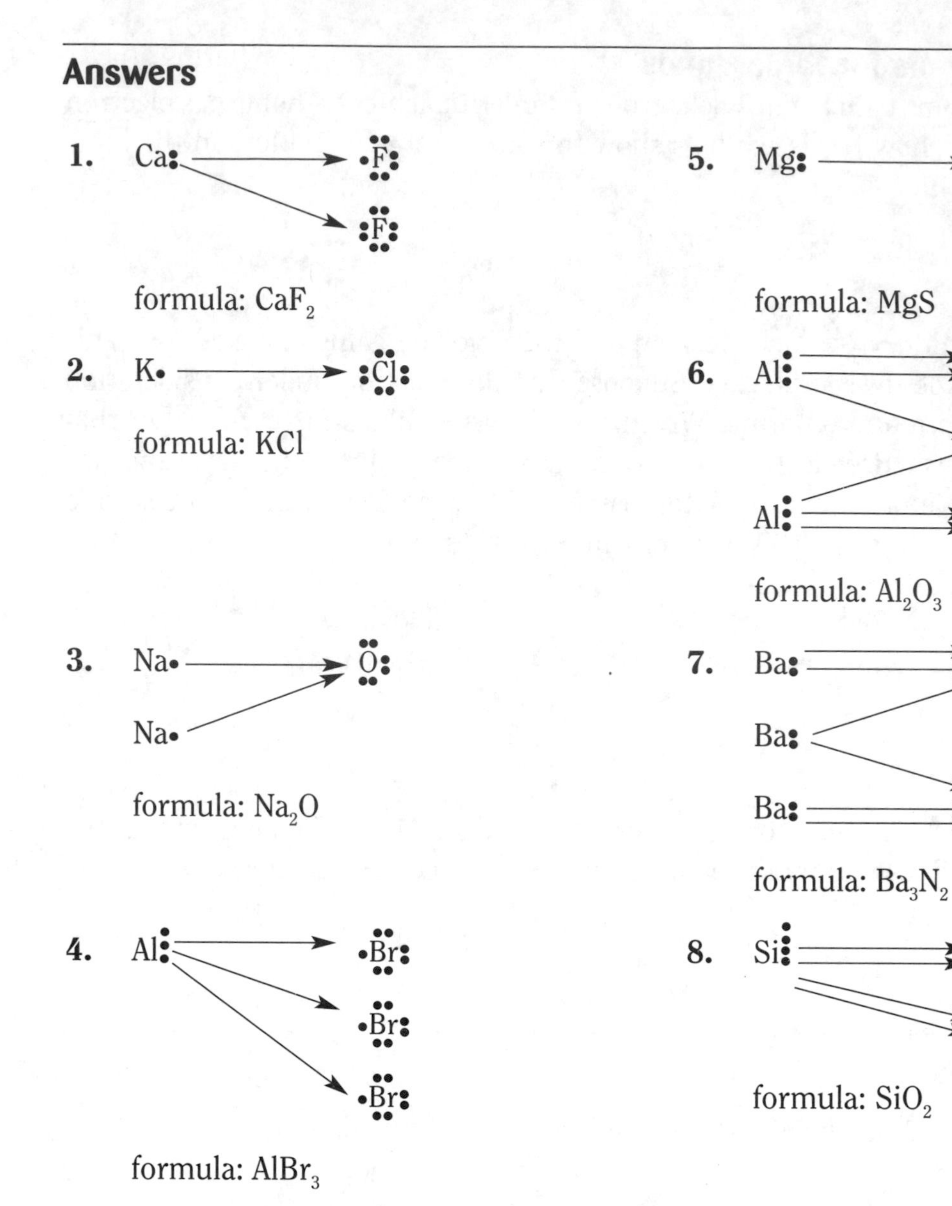

COVALENT BONDS

Covalent bonds are essentially bonds achieved by the *sharing* of electrons rather than by the transferring of electrons that characterizes ionic bonds. The octet rule is again in effect for the sharing of electrons. When placing valence electrons in the Lewis structure for the molecule in question, the shared electrons count as part of the octet for both atoms in the bond. For example, chlorine gas is a yellowish, toxic gas whose formula is Cl_2. Each of the chlorine atoms has seven valence electrons and so is in need of one more electron. Each chlorine atom shares one of the other chlorine's electrons, thereby giving each chlorine atom access to eight electrons, as shown in the following Lewis structure.

$$:\ddot{\underset{\cdot\cdot}{Cl}} \quad : \quad \ddot{\underset{\cdot\cdot}{Cl}}:$$

You can see that the increased stability that a chlorine atom achieves by sharing an electron with another chlorine atom is the reason why chlorine is a diatomic molecule, Cl_2. Covalent bonding is characterized by an electronegativity difference in the range of 0 to 1.3. This difference is small enough that neither of the bonding atoms is capable of pulling the shared electrons hard enough to create the electron transfer that leads to an ionic bond. Covalent bonds are further described as being either nonpolar or polar.

Bond Polarity

The idea of polarity is similar to the north and south poles of a magnet. However, in *bond polarity* we are talking about positive and negative rather than north and south. A bond has polarity when the shared pair of electrons is shared unequally. When one of the bonding atoms has more electron pulling power (more electronegativity) than the other atom, the shared pair of electrons is located closer to that atom. In the following illustration, suppose that X has an electronegativity of 1.8 and Y has an electronegativity of 1.0. Their shared pair of electrons is pulled closer to X. As electrons have a negative charge, this means that the X end of the bond is slightly more negative than the Y end of the bond. Chemists show this with what is called *delta negative and delta positive* notation, indicating a partial negative and a partial positive charge.

It is important to note that this is different from the entirely positive and entirely negative notation used for ions. It is a matter of degree.

Nonpolar Covalent Bonds

In a *nonpolar covalent bond*, the shared electrons are not drawn substantially closer to either of the two atoms. The situation is similar to a tug-of-war between nearly equally strong participants. Chemists agree that an electronegativity difference in the range of 0 to 0.2 results in a nonpolar covalent bond. A molecule having such a nonpolar bond is bromine gas (Br_2). Because each bromine atom has the same electronegativity, the difference between them is zero. Looking at the Lewis structure for bromine below, notice that the shared pair of electrons is not pulled to either end of the molecule, resulting in a nonpolar bond. There is no delta positive or delta negative notation because the electrons are uniformly distributed. We say that this bond has no polarity.

:Br : Br:

Polar Covalent Bonds

In a *polar covalent bond*, the shared electrons are shared unequally, being pulled more toward the atom with the higher electronegativity. As shown in Table 9.1, an electronegativity difference of 0.3 to 1.4 characterizes a polar covalent bond. This unequal electron sharing results in one end of the bond being slightly negative compared with the other end of the bond, which is slightly positive. In this case, the delta negative and delta positive notation can be used. In Figure 9.3, a bond between atom A, having an electronegativity of 2.1, and atom B, having an electronegativity of 1.5, has the shared electrons pulled closer to atom A.

A: shared pair of electrons B
(other electrons not shown)
2.1 1.5 (electronegativities)

Figure 9.3: Polar covalent bond

EXERCISES

In the following illustrations, the bonding electrons are pulled closer to the atom having the higher electronegativity. The greater the electronegativity difference, the more closely the electrons are pulled toward the atom with the greater electronegativity.

A :B C : D E: F

Examine the diagram in order to answer these questions.

1. Which of the letter combinations represents a nonpolar covalent bond?
2. Which of the letter combinations is most likely an ionic bond?
3. Which of the letter combinations is a polar covalent bond?

Answers

1. The CD bond is a nonpolar covalent bond because its shared pair of electrons is midway between the two atoms. This shows that atoms C and D have similar electronegativities, and so the difference is zero.
2. The AB bond is the best candidate for an ionic bond, as it appears that B has such high electronegativity that it has virtually captured A's bonding electron.
3. The EF bond is a polar covalent bond, as its shared pair of electrons is not at the midpoint, nor is it extremely close to either E or F.

Lewis Structures for Covalent Bonds

In drawing Lewis structures for covalently bonded molecules, the octet rule is used as a guide (remember that for hydrogen the octet is changed to a duet). Shared electrons count toward the electron total for both atoms in the bond. For example, the Lewis structure for SiH_4 looks like this:

H

H :Si: H

H

In creating this diagram, looking up hydrogen shows that it has one valence electron and that silicon has four. If silicon could share four more electrons, one from each hydrogen, it would have eight valence electrons. Hydrogen needs only two valence electrons for stability, so sharing one from silicon works for each hydrogen atom.

Another Lewis Structure

Yellowish, toxic chlorine gas (Cl_2) is called a *diatomic molecule*. Looking at its Lewis structure, you can see why it needs to be diatomic in order to have octet stability. This is an example of a nonpolar covalent bond because the electronegativity difference is zero.

:Cl : Cl:

Most chemists and most text books use a line to designate a shared pair of electrons. Redoing the two examples above produces

H
|
H — Si — H
|
H

:Cl —— Cl:

Double Bonds

A *double bond* is a covalent bond in which four electrons are shared rather than the two that are typical of a single covalent bond. As a practical matter, if you do not know that a double bond exists in a molecule, it soon becomes apparent that writing a Lewis structure having single bonds does not work. It is then that you should consider a double bond.

Example

Carbon dioxide

:Ö: :C: Ö: or   :Ö = C = Ö:

Triple Bonds

A *triple bond* is a covalent bond in which six electrons are shared. The nitrogen that is 80 percent of our atmosphere is N_2 and is an example of a molecule having a triple bond. Each nitrogen atom has five valence electrons, and so each needs to acquire another three electrons in order to satisfy the octet rule. Because both nitrogen atoms are in the same situation, a sharing of six total electrons works.

:N⋮   ⋮N: or   :N ≡ N:

Lewis Structures for Polyatomic Ions

In writing Lewis structures for polyatomic ions, follow these steps:

- Count all the valence electrons for the participating atoms.
- If the ion is a negative ion, add electrons to the total. Add one electron for a –1 ion, two electrons for a –2 ion, and so on.
- If the ion is a positive ion, subtract electrons from the total. Subtract one electron for a +1 ion, subtract two electrons for a +2 ion, and so on.
- Proceed to write the Lewis structure in the usual way.
- When finished, enclose the entire structure in brackets and place the charge of the ion at the upper right of the bracket.

Example

The sulfate ion (SO_4^{-2}) is found in car batteries, fertilizers, and epsom salts. Its Lewis structure is determined as follows.

The valence electron total is:

sulfur	= 6
oxygen	$4 \times 6 = 24$
ion charge	= 2
Total	= 32

Thus, there are 32 electrons to be placed in the Lewis structure:

Coordinate Covalent Bonds

A *coordinate covalent bond* is one in which both of shared electrons come from a single atom.

EXERCISES

Write the Lewis structures for each of the following:

1. ammonia (NH_3)
2. methane (CH_4)
3. sulfur dioxide (SO_2)
4. hydroxide ion (OH^-)
5. cyanide ion (CN^-)
6. carbon monoxide (CO)
7. sulfur trioxide (SO_3)
8. oxygen (O_2)
9. phosphorus trichloride (PCl_3)
10. hydrogen sulfide (H_2S)

Answers

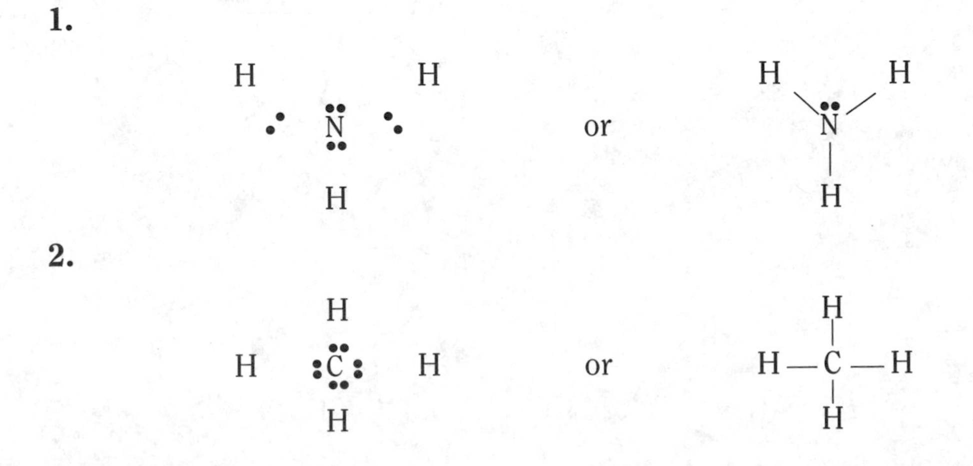

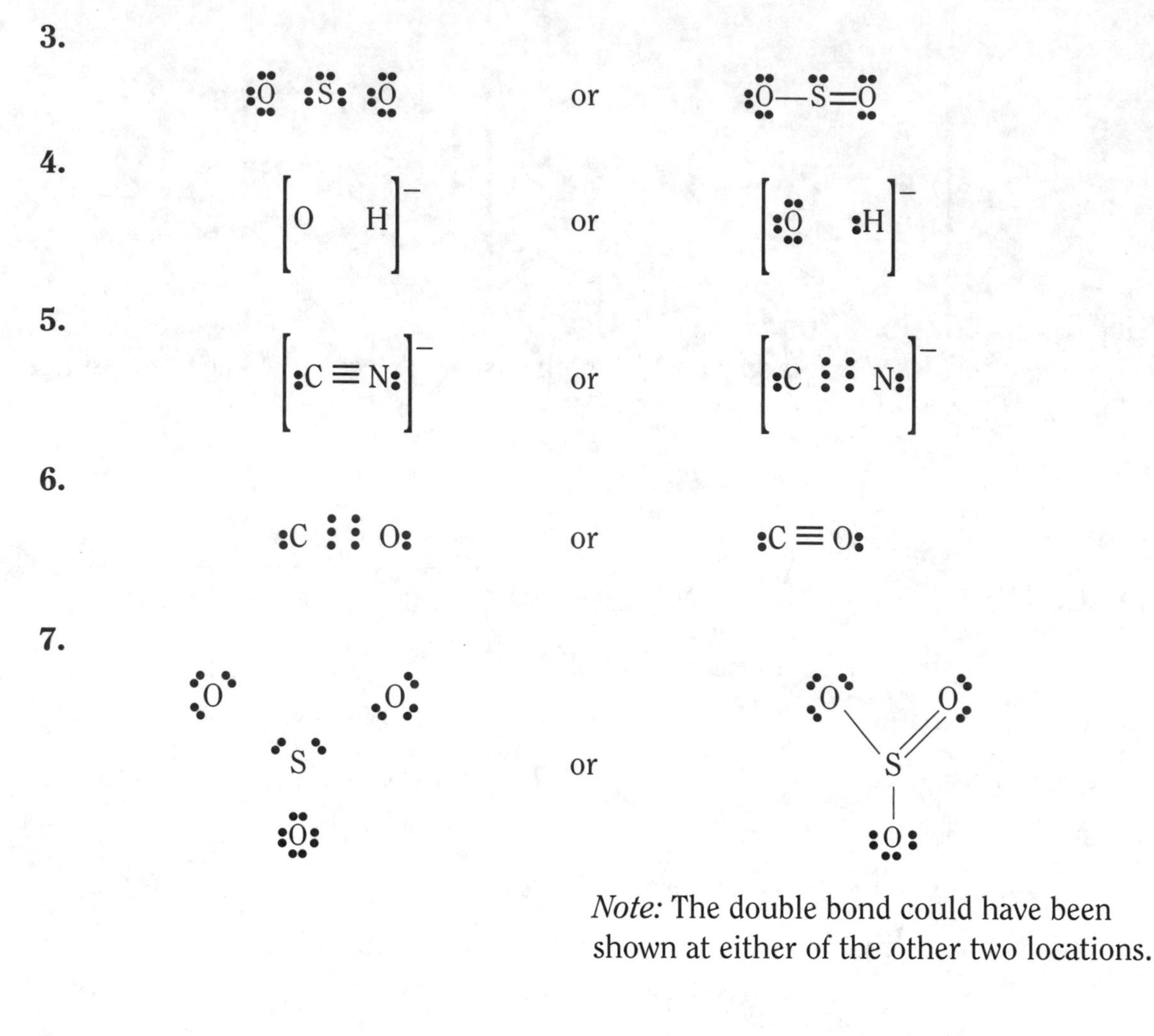

Note: The double bond could have been shown at either of the other two locations.

8.

:Ö : : Ö: or 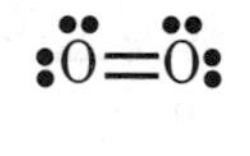

9.

:Cl: :Cl: P :Cl: or 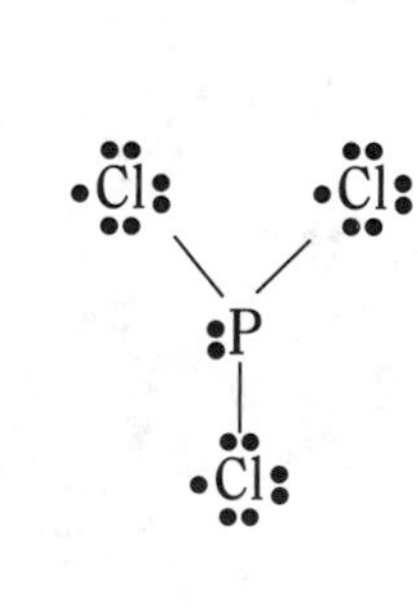

10.

H : Si : H or H—Si—H

HYDROGEN BONDING

A hydrogen bond is not a bond that holds a molecule together but rather is a bond between molecules. Some molecules that have hydrogen as one of their elements have bonds that are extremely polar. The significance of the hydrogen is that its small size allows the other element in the bond to pull the shared electrons very close. This extremely unequal sharing makes for a strongly polar bond. Water is the primary example of hydrogen bonding, and these bonds are illustrated by the dotted lines Figure 9.4.

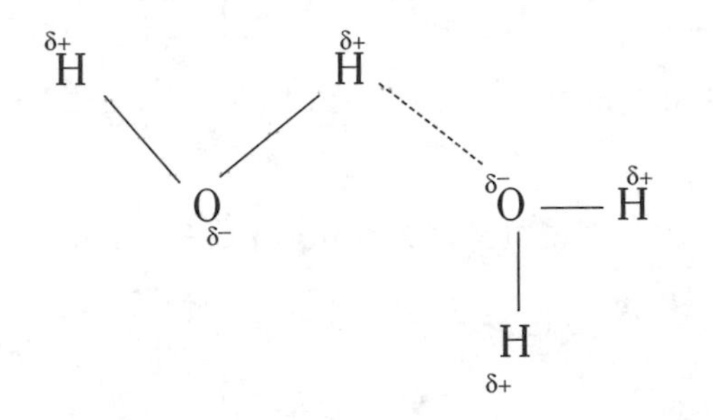

Figure 9.4: Hydrogen bonding in water

The fact that water molecules are usually held to each other by hydrogen bonds is responsible for the success of our planet and of life-forms themselves. Hydrogen bonding in water is the reason that frozen water is less dense than liquid water. Not only does ice float, but more importantly, floating ice permits water below the surface in ponds and lakes to remain in a liquid state during the winter and therefore allows for the continuity of life in these waters. On a more aesthetic note, hydrogen bonding is responsible for the six-sided nature of snowflakes.

A **hydrogen bond is the attraction that exists between the partially negative part of one molecule and the partially positive part of another molecule.** Hydrogen bonding is evident in water, ammonia (NH_3), and hydrofluoric acid (HF). The reason that hydrogen bonding is important is that it contributes to unusual properties in compounds that exhibit it. For example, the boiling point of these compounds is higher than would otherwise be expected. When a liquid boils, its molecules gain enough energy to break away from each other and become gaseous. Because hydrogen bonding causes molecules to be more strongly attracted to each other, a higher temperature is required for boiling. This relatively high boiling point for water ensures that our planet's bodies of water do not evaporate to extinction. It also makes our body chemistry possible at the temperatures of our environments.

EXERCISES

Answer the following questions about hydrogen bonding.

1. What is Lewis structure for HF?
2. Are the bonds polar or nonpolar?
3. Place the delta plus (δ^+) and delta minus (δ^-) at the correct places on the Lewis structure.
4. Write another Lewis structure for HF and place it next to your answer to question 1.
5. Using a dotted line, show the hydrogen bond that exists between the two molecules.

Answers

1. H:F: or H—F:
2. The bonds are polar because their electronegativity difference is greater than 0.2.
3. $\overset{\delta+}{H}$ — $\overset{\delta-}{F}$:
4. $\overset{\delta+}{H}$ — $\overset{\delta-}{F}$: $\overset{\delta+}{H}$ — $\overset{\delta-}{F}$:
5. $\overset{\delta+}{H}$ — $\overset{\delta-}{F}$: ------------------- $\overset{\delta+}{H}$ — $\overset{\delta-}{F}$:

METALLIC BONDS

The atoms of metals exhibit **metallic bonding**, in that the valence electrons of the metal atoms form a pool of electrons. This pool of electrons is shared by all the atoms and therefore serves to knit them together. Because these electrons are free to move, metals are excellent conductors of electricity. Other qualities of metals, including *malleability*, the ability to be hammered into forms, and *ductility*, the ability to be drawn into wires are facilitated by the flexibility of this sea of electrons.

Resonance

There are times when the Lewis structures written for a molecule or ion do not adequately explain its properties. Ozone (O_3) is just such an example. The Lewis structure is

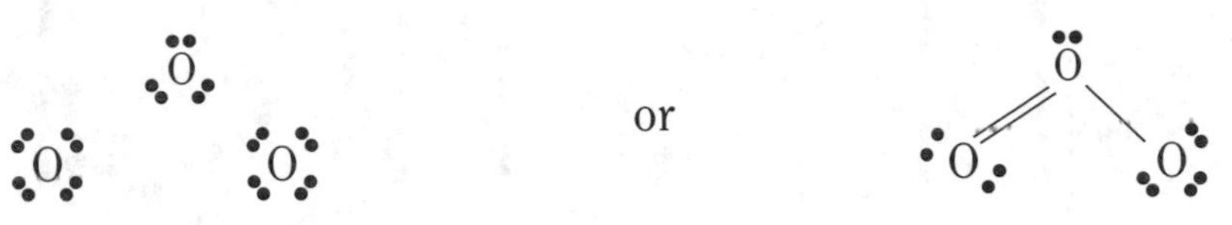

However, this diagram makes it appear that ozone has two different kinds of bonds, single and double. In fact, experimental data shows that ozone has two identical bonds. At this point in the reasoning, if we write the other Lewis possibility for ozone, it will be

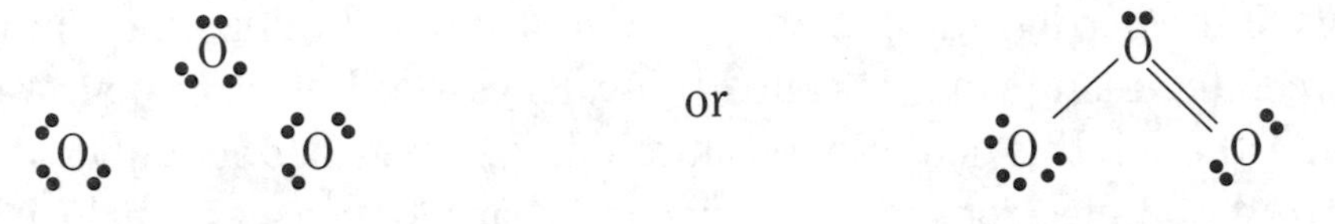

These two equivalent Lewis structures are called **resonance forms**. In describing the structure of ozone, we draw both structures and say that ozone is represented by an average of the two structures:

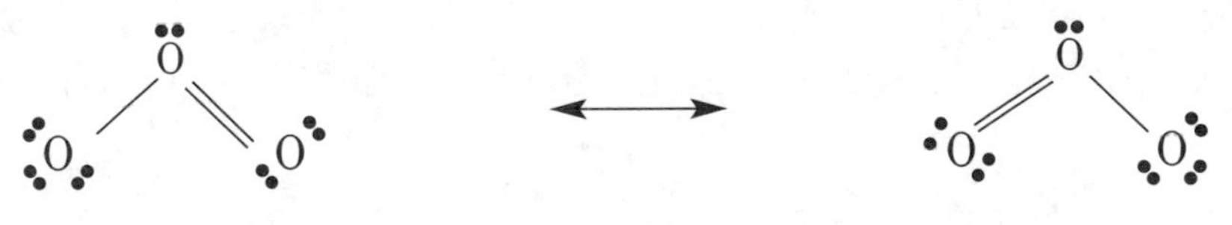

The double arrow indicates that these Lewis structures are resonance forms. It does not mean that the molecule vibrates back and forth between these two forms but rather that there is just one form of ozone. The reason for writing the two structures is that there is a limitation in the ability of Lewis structures to describe the electron distribution in some molecules.

EXERCISE

Write all the possible Lewis structures for the nitrate ion (NO_3^-). Experimental data shows that the nitrate ion has three identical bonds. Show that the structures are resonance forms by the correct use of arrows.

Answer

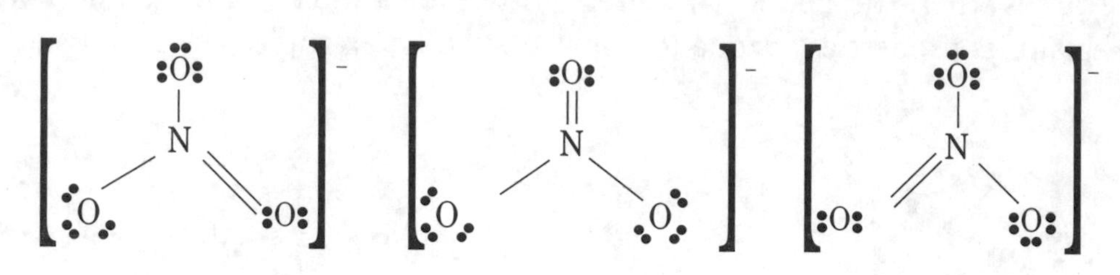

MOLECULAR SHAPES

The shape of a molecule is a function of the interactions among the valence electrons in the outer orbitals of the participating atoms in the molecule. The shapes of simple molecules can be predicted using what is called the **valence-shell electron-pair repulsion (VSEPR) model.** Although this name sounds forbidding, the idea is actually straightforward. Remembering that electrons are negative and that like charges repel each other, you can imagine that the outermost (valence) electrons in a molecule push away from each other as far as possible. In the VSEPR model this pushing is called *electron pair repulsion.* The shape of a molecule is whatever shape allows the electron pairs to be as far from each other as possible.

To see how the VSEPR model works, examine the methane (CH_4) molecule. The first step is to write its Lewis structure.

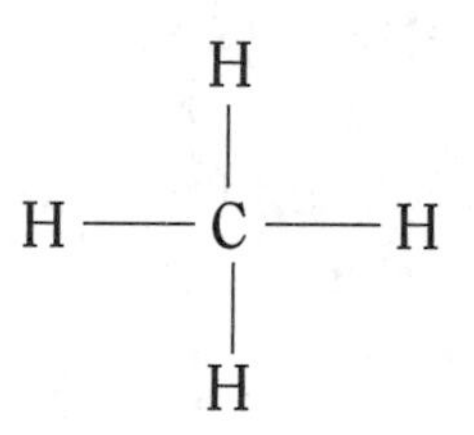

In order for the electron pairs to be as far from each other as possible in a three-dimensional space, the molecule must have a tetrahedral shape, as shown in Figure 9.5.

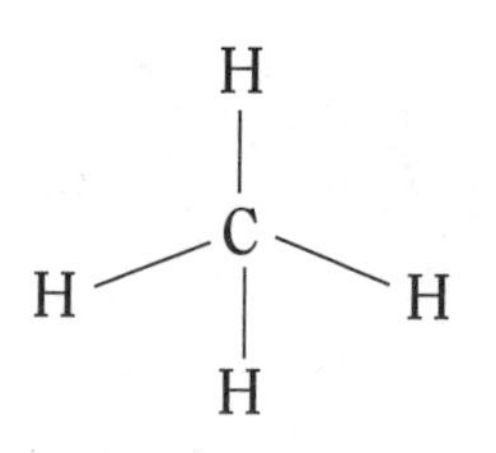

Figure 9.5: Tetrahedron

The idea of a tetrahedral shape is not confined to molecules having four atoms bonded to a central atom. An examination of the ammonia (NH_3) molecule reveals another type. However, in this molecule the groups of valence electrons are in a tetrahedral configuration, but the geometry of the molecule itself is pyramidal. The first step in understanding this idea is to write its Lewis structure:

$$\begin{array}{c} H - \ddot{N} - H \\ \\ H \end{array}$$

As you look at this Lewis structure, notice that there are four pairs of electrons. There are three shared pairs, denoted by the lines, and one unshared pair, represented by the dots above the N atom. The unshared pair is also called a *lone pair*. Ammonia's four pairs of electrons are all valence electrons. The shape that allows these four pairs of electrons to be as far from each other as possible places them at the corners of a tetrahedron, as shown in Figure 9.6. This is called *electron group geometry*. The arrangement of the atoms is called *molecular geometry*, which in this case is pyramidal.

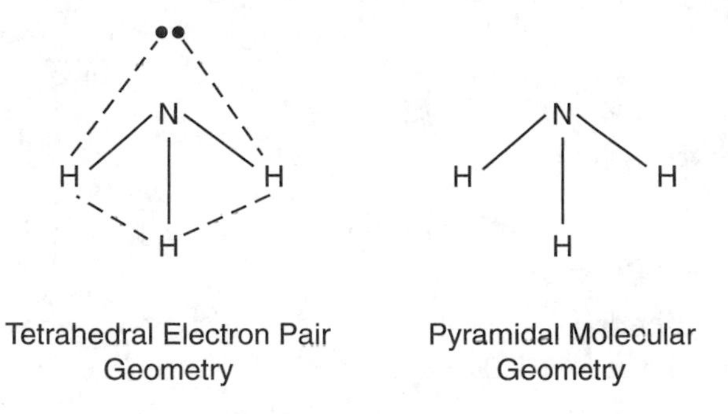

Figure 9.6: Electron group geometry

In Figure 9.6, a diagram of the ammonia molecule, notice that the **molecular geometry** is pyramidal because that is how its atoms are arranged in space. However, its **electron pair geometry** is tetrahedral and it is the electron pair geometry that dictates the molecular geometry.

The steps involved in determining the shape of a molecule are as follows.

- Write the Lewis structure of the molecule.
- Count the total number of electron pairs around the central atom.
- Count the number of lone pairs of electrons around the central atom.
- Use the number of electron groups, the number of lone pairs of electrons, and Table 9.2 to determine both the electron group geometry and the molecular geometry.

TABLE 9.2: ELECTRON AND MOLECULAR GROUP GEOMETRY

No. of Electron Groups	Electron Group Geometry	No. of Lone Pairs	Molecular Geometry*	Molecular Shape
Two	linear	none		linear
Three	trigonal planar	none	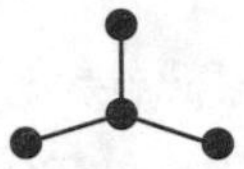	trigonal planar
Three	angular	one	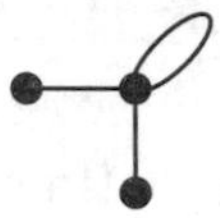	angular
Four	tetrahedral	none	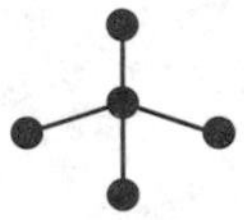	tetrahedral
Four	tetrahedral	one		pyramidal
Four	tetrahedral	two		angular

* 0 indicates a lone electron pair.

Examples

1. Determine the shape of the water molecule (H_2O). Rewrite its Lewis structure using the correct shape.

   Step 1. Write the Lewis structure:

   $$H - \ddot{\underset{\cdot\cdot}{O}} - H$$

   Step 2. Count the electron groups: four

   Count the lone pairs two

   Step 3. This results in an angular molecule as shown in Table 9.2.

   The electron group geometry is tetrahedral.

Step 4. Rewrite the Lewis structure using the correct shape.

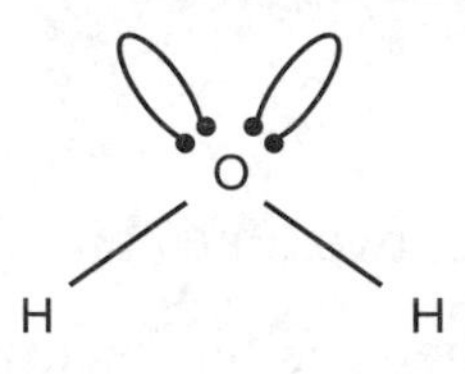

2. Determine the shape of the PF_3 molecule.

Step 1. Write the Lewis structure:

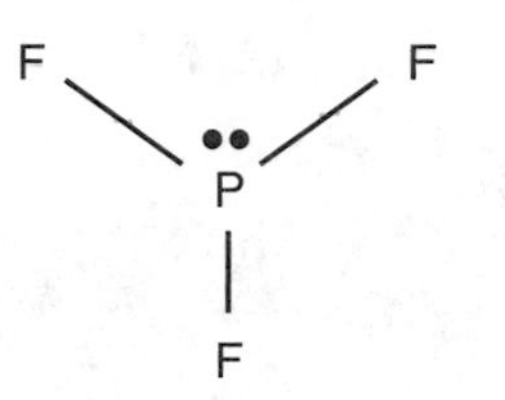

Step 2. Number of electron groups: four

Number of lone pairs: one

Step 3. Molecular shape: trigonal pyramid

Electron group geometry: tetrahedral

Step 4. Rewrite the Lewis structure using the correct shape.

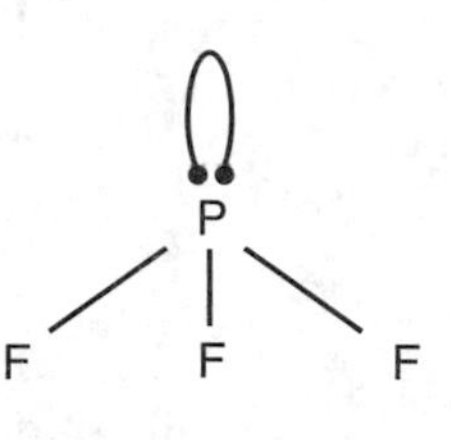

EXERCISES

For each of the following, write the Lewis structure, identifying the number of electron groups and the number of lone pairs. Determine the electron group geometry and the molecular geometry. Rewrite the Lewis structure in the correct shape.

1. hydrogen sulfide (H_2S)
2. carbon tetrachloride (CCl_4)
3. phosphorus trihydride (PH_3)

Answers

1. Step 1. $H - \ddot{\underset{\cdot\cdot}{S}} - H$

Step 2. Four electron groups, two lone pairs

Step 3. Electron group geometry is tetrahedral; molecular geometry is angular.

Step 4.

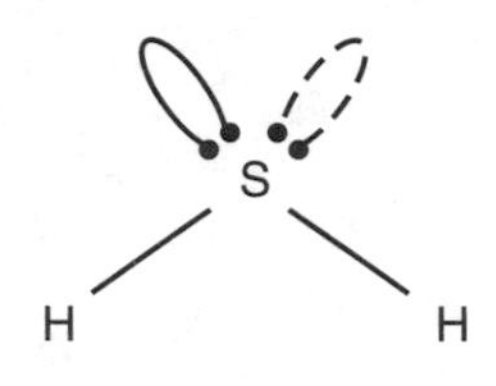

2. Step 1.

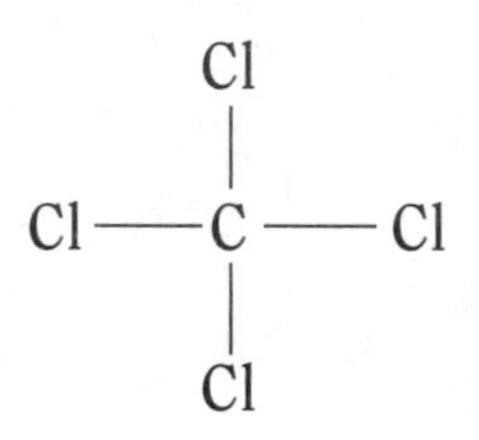

Step 2. Four electron groups, no lone pairs

Step 3. Electron group geometry is tetrahedral; molecular geometry is tetrahedral.

Step 4.

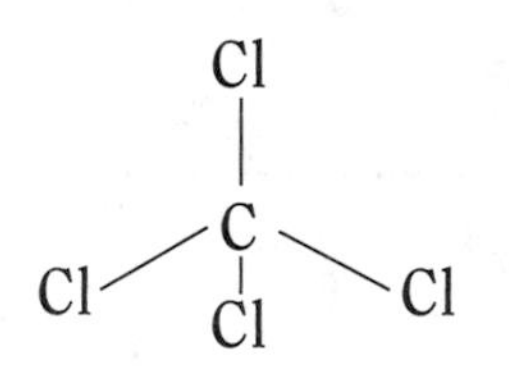

3. Step 1.

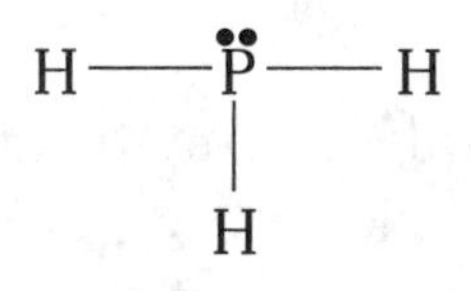

Step 2. Four electron groups, one lone pair

Step 3. Electron group geometry is tetrahedral; molecular geometry is pyramidal.

Step 4.

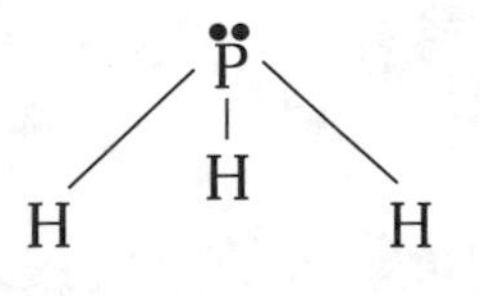

BONDING IMPLICATIONS

How atoms are bonded to each other plays a significant role in the properties of the substances they create. Saturated and unsaturated fats are listed on the nutritional labels of many food products, and is medical advice about the advisability of including them in a healthy diet is offered. *Saturated fats* have single bonds, whereas *unsaturated fats* have double bonds. The triple bond in acetylene is responsible for its extreme reactivity and flammability.

The strength of ionic bonds is responsible for high melting points. For example, it is impossible to melt table salt (NaCl) without industrial-strength heating equipment. On the other hand, covalently bonded sugar ($C_{12}H_{22}O_{11}$) can be melted in the broiler unit of an ordinary stove. Solutions of ionic compounds carry an electric current, whereas solutions of covalently bonded molecules do not. As you continue your study of chemistry, you will begin to look at reactions and properties as functions of chemical bonds.

UNIT 10

Oxidation and Reduction

Electron Gainers and Losers, Oxidizing and Reducing Agents, Electrochemistry

Oxidation and reduction are two sides of the same coin, in that *oxidation* is about the loss of electrons and *reduction* is about the gain of electrons. Within any chemical reaction, if oxidation is occurring, reduction is also taking place. This simultaneous occurrence of oxidation and reduction is called *redox*, a combination of the words *red*uction and *ox*idation. The redox reactions of our everyday life include respiration, rusting, all combustion, and most metabolic reactions in living organisms.

In order to examine the workings of oxidation and reduction, the first order of business is to determine the *oxidation number*. The oxidation number is essentially what we have called charge so far. Up to now we have simply looked up the charge in a table, but now we have a slightly different set of rules for determining the oxidation number, as there is some variation not predicted by tables.

OXIDATION NUMBER RULES

- The oxidation number of an atom in a pure element is 0.
- Oxygen has an oxidation number of –2. (There are exceptions, but we will not consider them here.)
- Elements in the first column of the periodic table have an oxidation number of +1.
- Elements in the second column of the periodic table have an oxidation number of +2.

- The oxidation number of other elements can be determined by using the strategy shown below.

How to Calculate Oxidation Numbers Not Covered by Rules

The sum of all the oxidation numbers in a formula must equal zero. By using the oxidation numbers that you know from the oxidation number rules, combined with the idea of a zero total you can calculate the oxidation number of an element that does not fit the rules.

Examples

Determine the oxidation number of the underlined element in each of the following.

1. $\underline{S}O_2$

$S + 2(-2) = 0$

$S - 4 = 0$

$S = +4$

2. $K_2\underline{S}O_3$

$2(+1) + S + 3(-2) = 0$

$2 + S - 6 = 0$

$S = 6 - 2 = +4$

If the element in question has a subscript in the formula, such as the 2 in Fe_2O_3, the solution is as follows.

3. $\underline{Fe_2}O_3$

$2(Fe) + 3(-2) = 0$

$2Fe - 6 = 0$

$2Fe = +6$

$Fe = +3$

If the subscript is in an elemental formula, such as O in O_2 or in O_3, the single oxygen atom has an oxidation number of 0. (See the first rule.)

EXERCISE

Determine the oxidation number of the underlined element in each of the following.

1. $\underline{C}O_3$
2. $\underline{C}$
3. $\underline{C}O$
4. $H_2\underline{S}O_4$
5. $K_2\underline{Cr}O_4$
6. $\underline{Mg}$
7. $H\underline{Sb}F_6$
8. $\underline{Al}_2O_3$

Answers

1. +4
2. 0
3. +2
4. +6
5. +6
6. 0
7. +5
8. +3

There will be times when you will need to determine the oxidation number of an element in a polyatomic ion, for example, Mn in MnO_4^{-1}. The only change in strategy is that the sum of the oxidation numbers must equal the charge of the polyatomic ion rather than the zero used previously.

$$\begin{array}{llcl} Mn & +4(-2) & = & -1 \\ Mn & -8 & = & -1 \\ Mn & & = & -1 + 8 = +7 \end{array}$$

Here is an example of another polyatomic ion.

$$\begin{array}{llcl} \underline{Cr}_2O_7^{-2} & & & \\ 2Cr & +7(-2) & = & -2 \\ 2Cr & & = & -2 + 14 \\ 2Cr & & = & +12 \\ Cr & & = & +6 \end{array}$$

EXERCISES

Determine the oxidation number for the underlined element in each of these polyatomic ions.

1. $\underline{N}O_3^-$
2. $\underline{P}O_4^{-3}$
3. $\underline{C}O_3^{-2}$
4. $\underline{I}O_3^{-1}$
5. $\underline{S}_2O_3^{-2}$

Answers

1. +5
2. +5
3. +4
4. +5
5. +2

What to Do With Oxidation Numbers

Now that you can determine oxidation numbers, what do you do with them? Oxidation and its companion, reduction, involve the gain and loss of electrons. If you know the oxidation number of an element before and after it reacts, you can tell what has happened to its electrons and therefore which redox process has taken place. For example, suppose that oxygen goes from its elemental zero state to its –2 state:

$$O^0 \longrightarrow O^{-2}$$

Is this a gain or a loss of electrons? Recalling that electrons are negative particles reveals that two electrons are gained. *Reduction is a gain of electrons.* It helps to remember that it is the oxidation number that is reduced, in this case from 0 down to –2.

Oxidation is a loss of electrons. For example, the oxidation number increases when elemental iron loses electrons in the rusting process to become the +3 ion:

$$Fe^0 \longrightarrow Fe^{+3}$$

It is a bit harder to see this iron transformation as being a loss of electrons. In Fe^0, the number of electrons equals the number of protons, resulting in a zero oxidation number. In Fe^{3+} you can tell that the positive protons outnumber the negative electrons by 3. Recalling that all forms of any element have the same number of protons, it must be that three electrons have been lost, resulting in the protons being ahead by 3. Strangely enough, oxidation does not necessarily involve oxygen.

$$2Al + 3I_2 \longrightarrow 2AlI_3$$

The aluminum oxidizes from its zero elemental state to its +3 state, whereas the iodine is reduced from its zero elemental state to its –1 state.

Here are some examples of oxidation and reduction.

$Mg^0 \longrightarrow Mg^{2+}$	oxidation: loss of two electrons
$Cl^0 \longrightarrow Cl^-$	reduction: gain of one electron
$Mn^{4+} \longrightarrow Mn^{2+}$	reduction: gain of two electrons
$O^{-2} \longrightarrow O^0$	oxidation: loss of two electrons

EXERCISES

Decide if the following changes represent oxidation or reduction.

1. $Fe^{2+} \longrightarrow Fe^{3+}$
2. $H^0 \longrightarrow H^+$
3. $Al^{3+} \longrightarrow Al^0$
4. $Ca^0 \longrightarrow Ca^{2+}$
5. $S^0 \longrightarrow S^{2-}$

Answers

1. oxidation
2. oxidation
3. reduction
4. oxidation
5. reduction

SPECTATOR IONS

A *spectator ion* is one whose oxidation number does not change. That is, its oxidation number on the reactant side of the equation is the same as its oxidation number on the product side of the equation. In short, spectator ions are like other spectators, present but not actively involved.

Examples

$$2KClO_3 \longrightarrow 2KCl + 3O_2$$

After determination of the oxidation numbers, we have

$$2K^{+}Cl^{5+}O_3^{2-} \longrightarrow 2K^{+}Cl^{1-} + 3O_2^{0}$$

The K^+ ion is the spectator because it has a +1 oxidation number on both sides of the equation.

Spectator ions can be single ions, such as the K^+ ion in the previous example, or they can be polyatomic ions, such as nitrate, sulfate, and so on.

Example

$$Cu + 2AgNO_3 \longrightarrow 2Ag + Cu(NO_3)_2$$

$$Cu^0 + 2Ag^{+}NO_3^{-} \longrightarrow 2Ag^0 + Cu^{2+}(NO_3)^{-}_2$$

The nitrate ion (NO_3^{-1}) is the spectator ion with a –1 oxidation number throughout.

LOOKING AT WHOLE EQUATIONS

Up to this point we have evaluated parts of reactions. Now we will look at the entire reaction, with the object being to write the oxidation part and the reduction part separately. These parts are called *half-cell* reactions. In order to do this:

- Determine the oxidation number for each element on both sides of the equation.
- Examine these numbers to see which element was oxidized and which element was reduced. Also, look to see if there are one or more spectator ions.

- Write the first half-cell reaction, showing the species being oxidized first in its reactant state and then in its product state.
- Write the other half-cell reaction, showing the species being reduced first in its reactant state and then in its product state.
- Note any spectator ion(s) if present.

Example

$2Al + Fe_2O_3 \longrightarrow 2Fe + Al_2O_3$

$2Al^0 + Fe^{3+}{}_2O^{2-}{}_3 \longrightarrow 2Fe^0 + Al^{3+}{}_2O^{2-}{}_3$

oxidation: $Al^0 \longrightarrow Al^{+3}$

reduction: $Fe^{3+} \longrightarrow Fe^0$

spectator: O^{2-}

EXERCISES

After determining the oxidation numbers of each species, write the half-cell reactions. If there is a spectator ion, identify it.

1. $Mg + Cu(NO_3)_2 \longrightarrow Cu + Mg(NO_3)_2$
2. $C + O_2 \longrightarrow CO_2$
3. $2AgBr \longrightarrow 2Ag + Br_2$
4. $2SO_2 + O_2 \longrightarrow 2SO_3$
5. $Zn + 2HCl \longrightarrow ZnCl_2 + H_2$

Answers

1. $Mg^0 \longrightarrow Mg^{2+}$ oxidation

   $Cu^{2+} \longrightarrow Cu^0$ reduction

   $NO_3{}^{1-} \longrightarrow$ spectator

2. $C^0 \longrightarrow C^{4+}$ oxidation
   $O^0 \longrightarrow O^{2-}$ reduction
3. $Ag^+ \longrightarrow Ag^0$ reduction
   $Br^- \longrightarrow Br^0$ oxidation
4. $S^{4+} \longrightarrow S^{6+}$ oxidation
   $O^0 \longrightarrow O^{2-}$ reduction
5. $Zn^0 \longrightarrow Zn^{2+}$ oxidation
   $H^+ \longrightarrow H^0$ reduction
   $Cl^{1-} \longrightarrow$ spectator

CHEMICAL AGENTS

Agents cause a chemical reaction. An *oxidizing agent* is a chemical species that causes oxidation. Because oxidation is a loss of electrons, an oxidizing agent must be a species (molecule, ion, or atom) that pulls electrons toward itself. As an oxidizing agent acquires electrons, it is reduced. A *reducing agent* is a chemical species that causes reduction. Although this sounds twisted, perhaps the following table will help.

Oxidizing Agent	Reducing Agent
Facilitates oxidation	facilitates reduction
Accepts electrons	donates electrons
Is reduced	is oxidized

EXERCISES

Consider the reaction

$$A^{3+} + B^0 \longrightarrow A^0 + B^+$$

1. Which species is oxidized?
2. Which species is reduced?
3. Which is the oxidizing agent?
4. Which is the reducing agent?
5. The charge of which species is decreasing?
6. Is there a spectator ion?

Answers

1. B^0
2. A^{3+}
3. A^{3+}
4. B^0
5. A^{3+}
6. no

ACTIVITY SERIES

As you learned in the previous units on bonding and periodicity, metals are losers of electrons. They do not all lose electrons with equal ability, but rather some are better losers than are others. The ease or relative difficulty with which metals lose electrons is related to their electron configurations and their atomic sizes. The *activity series* (Table 10.1) is a listing of some commonly used metals arranged in order of how easily they lose electrons. The typical heading for this listing is "strength as a reducing agent," which we know means how well a metal loses electrons.

TABLE 10.1: ACTIVITY SERIES

Ability to Act as a Reducing Agent	Element
Powerful	K (potassium) Na (sodium) Ca (calcium)
Strong	Mg (magnesium) Al (aluminum) Cr (chromium)
Good	Zn (zinc) Fe (iron) Cd (cadmium)
Fair	Ni (nickel) Sn (tin) Pb (lead)
Poor	Cu (copper) Ag (silver) Hg (mercury)
Very poor	Au (gold)

EXERCISES

1. Magnesium is a shiny, silvery metal. Why is it not suitable for use in jewelry?
2. What is it about lead that makes it a better candidate than iron for use in plumbing?
3. Is gold oxide likely to form on gold jewelry?

Answers

1. Magnesium is not suitable for jewelry because it loses electrons too easily. (It is a strong oxidizing agent.) As it loses electrons, it loses its metallic strength, becoming an ion.
2. Lead loses electrons less readily than does iron. (It is a weaker reducing agent.) This allows lead to keep its metallic strength better.
3. Gold oxide is not likely to form on gold jewelry because gold (Au) is a very poor loser of electrons. In order to combine with oxygen, an element must be able to lose electrons.

ELECTROCHEMISTRY

The gain and loss of electrons in oxidation and reduction reactions can be used to create an electric current. A *galvanic cell* is an arrangement of materials such that an electric current is produced. Examine the schematic diagram of a galvanic cell in Figure 10.1.

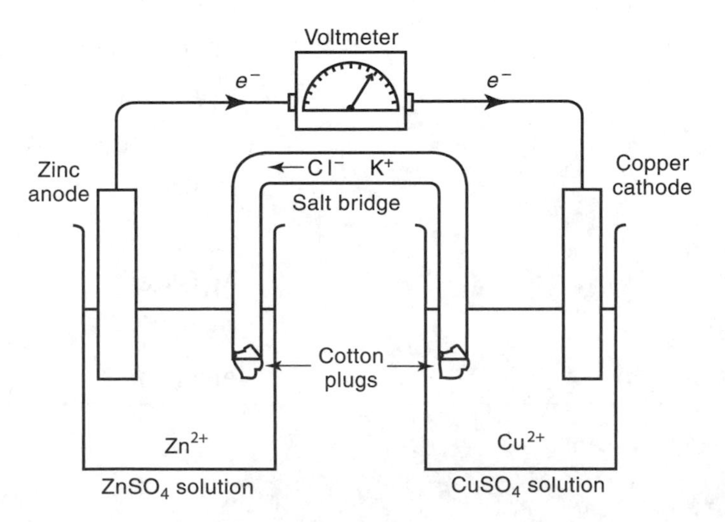

Figure 10.1: Galvanic cell

The half-cell reactions are

$$Zn^0 \longrightarrow Zn^{2+}$$

$$Cu^{2+} \longrightarrow Cu^0$$

The activity series shows that zinc is a better loser of electrons than copper. It is this relative difference in the two metals that causes zinc to be the electron loser in the zinc/copper cell. Zinc responds differently when paired with a metal higher that it is in the activity series.

Example

Which of the half-cell reactions in the galvanic cell in Figure 10.1 is oxidation? (Zn^0 becoming Zn^{2+} is oxidation.) The two metal strips, Zn^0 and the Cu^0, are called *electrodes*. The electrode where oxidation takes place is called the *anode*. The electrode where reduction takes place is called the *cathode*. In Figure 10.1, which electrode is the anode? (The Zn^0 is the anode.)

A way to remember the electrode names is that anode and oxidation both begin with a vowel, and cathode and reduction both begin with a consonant.

Continuing with the examination of the electrochemical cell, notice that electrons are lost by the zinc side of the cell and taken in by the copper side of the cell. For this reason, electricity flows from left to right in the diagram.

Salt Bridge

A *salt bridge* is a glass tube that contains a salt solution and has semiporous plugs that allow the salt's ions to migrate into the solution. As you look at the half-cell reactions, you can see that the zinc reaction produces positive ions. For this reason, the zinc beaker would become more and more positive if not for the negative ions from the salt bridge migrating into this beaker. The positive ions from the salt bridge migrate to the copper beaker because it is losing its Cu^{2+} ions. Without the salt bridge, electricity would not flow.

EXERCISES

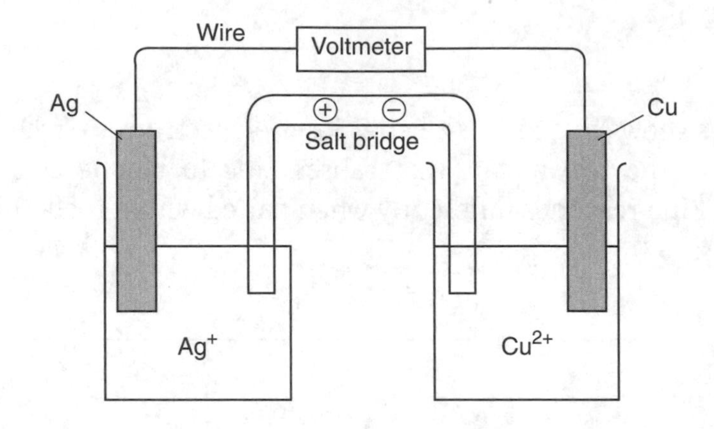

From Table 10.1, you can tell that in Figure 10.2 copper is a better loser of electrons than is silver. This knowledge allows you to decide which element participates in oxidation and which one participates in reduction in the equation.

1. Write the two half-cell reactions.
2. Is oxidation taking place at the silver or the copper electrode?
3. Does the amount of Cu^{2+} increase, decrease, or remain the same?
4. Which electrode is the cathode?
5. Does electricity flow from left to right or from right to left?
6. Into which beaker do positive ions from the salt bridge migrate?
7. If there were no salt bridge, what would happen?

Answers

1. Cu^0 $\quad$ Cu^{2+}

   Ag^+ $\quad$ Ag^0
2. Oxidation is taking place at the copper electrode.
3. The number of Cu^{2+} ions increases, as they are being produced by the oxidation.
4. The cathode is Ag^0 because reduction is occurring there.
5. Right to left, because the Cu^0 electrode is losing electrons, and they are being used on the left in the Ag^+ beaker.

6. The positive ions from the salt bridge migrate to the silver beaker to replace the positive silver ions being used up by the reaction.
7. Without a salt bridge, no electricity would flow.

BATTERIES

Batteries are a practical application of the galvanic cell in that an oxidation–reduction reaction generates an electric current. A battery that has an enormous impact on our lives is the automobile battery, shown in Figure 10.3.

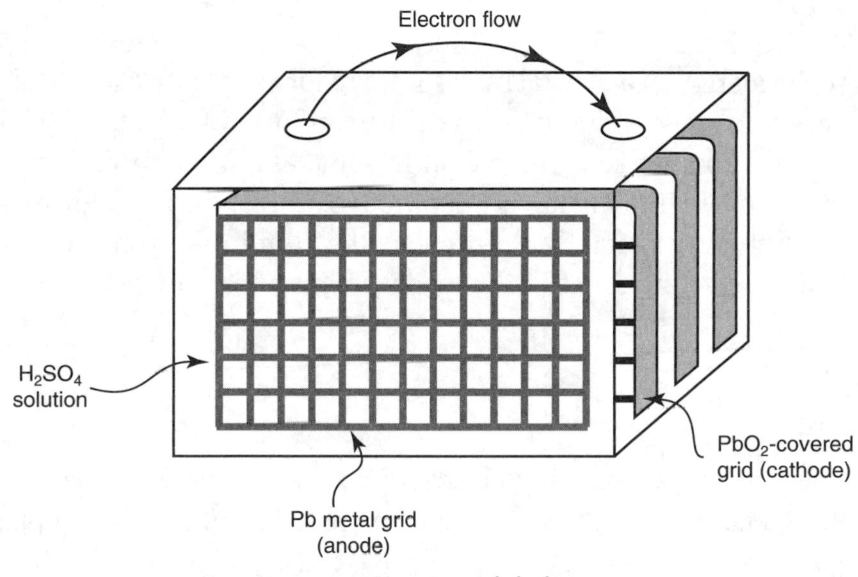

Figure 10.3: Automobile battery

The overall reaction in an automobile battery is

$$Pb + PbO_2 + 2H_2SO_4 \longrightarrow H_2O + 2PbSO_4$$

The lead loses electrons as it oxidizes:

$$Pb \longrightarrow Pb^{2+}$$

The lead +4 ion in the PbO_2 gains electrons as it is reduced.

$$Pb^{4+} \longrightarrow Pb^{2+}$$

The sulfuric acid (H_2SO_4) is used up by the battery's reaction.

The water produced by the battery's reaction is a cause of potential trouble in the jump-starting of a dead battery. The electricity provided by the good battery used in the jump-starting can cause electrolysis of water according to this reaction:

$$2H_2O \longrightarrow 2H_2 + O_2$$

The combination of elemental hydrogen (H_2) and elemental oxygen (O_2) is a potentially explosive mixture that can be ignited by a stray spark.

CORROSION

Corrosion is the oxidation of a neutral metal atom to form a positive ion. The reason that this is not a desirable reaction is that the positive ion form of a metal is generally soluble and so the corrosion weakens the metal. In some cases, the oxidation actually creates a protective film over the metal. As aluminum oxidizes, the Al_2O_3 formed adheres so strongly to the aluminum metal that it protects the aluminum from further oxidation. Millions of dollars a year are spent in the management of corrosion in the home and in industry.

Electroplating

Electroplating is the process of using electricity to create a very thin layer of metal on an object such as a car bumper, a ring, or dinnerware. Essentially electroplating is

$$M^+ + \text{electricity} \longrightarrow M^0$$

The object to be plated is suspended in a solution of positive ions (gold, chromium, silver). When an electric current is provided to the object being plated, the positive ions in the solution take up the electrons being provided by the current and become neutral metal atoms. These atoms then attach themselves to the object being plated. In the case of jewelry, the object being plated is made of a heavy metal so that the finished product will have the heft or feel of gold or silver. Electroplating is a means of using an ultrathin metal coating for applications that would be prohibitively expensive otherwise.

UNIT 11

Acids and Bases

The Roles of the Hydronium and Hydroxide Ions, pH, and Neutralization Reactions

Acids, and to a lesser extent bases, are part of our conversations, advertisements, and concerns. Headlines and television screens inform and persuade us about acid reflux, acid rain, cleaning compounds, and antacids (Figure 11.1). As you work your way through this unit and learn about acids and bases, you will find yourself looking at their presence in the world around you with a new awareness.

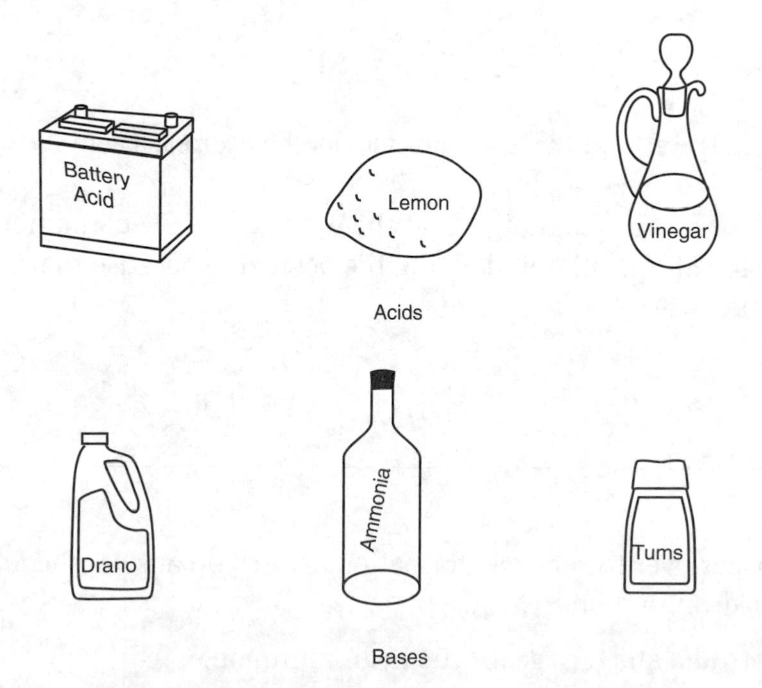

Figure 11.1: Common uses for acids and bases

Acids are defined as compounds that are hydrogen ion donors. They are also sometimes referred to as proton donors because a hydrogen ion is really just a proton. For an acid to be a hydrogen ion donor, its formula must contain at least one hydrogen atom. However, there are compounds that have hydrogen atoms but are not acids. For example, water (H_2O), wood alcohol (CH_3OH), and lye ($NaOH$) all have hydrogen atoms but are not acids. This difficulty led to a fine-tuning of the definition of acids:

***Acids* are compounds that produce hydronium ions when mixed with water.**

THE HYDRONIUM ION

The hydronium ion is a very special ion because it is *the* defining particle of all acids. A hydrogen ion is so reactive that it really does not exist by itself, but rather it attaches to something else, usually a water molecule. When a hydrogen ion (H^+) attaches itself to a water molecule, the resulting combination is called a *hydronium ion*.

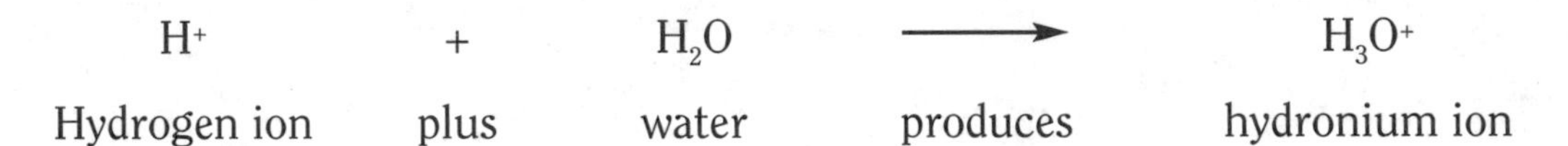

Hint: The hydronium ion is the focus of all acid–base studies, so keep a picture of it in your mind (Figure 11.2).

$$\left[\begin{array}{c} H - O - H \\ H \end{array}\right]^+$$

Figure 11.2: The Lewis structure for the hydronium ion

As you examine Figure 11.2, notice that the water molecule could not have accepted a whole hydrogen atom without violating the octet rule because there would be one too many electrons.

EXERCISES

Place a piece of paper over the above information before you answer the following questions about the hydronium ion.

1. Write the formula and charge for the hydronium ion.
2. What does a water molecule acquire to produce a hydronium ion?

3. What ion is typical of the reaction of all acids with water?
4. A company producing bottled water chose to name its product "Extreme Water H_3O^+." Why would this name have no appeal for someone who knows chemistry?

Answers

1. H_3O^+
2. a hydrogen ion
3. hydronium (H_3O^+)
4. It would seem that the water is really an acid.

IT'S AN ACID IF . . .

All acids produce a hydronium ion when combined with water. The acid donates a hydrogen ion, which attaches itself to a water molecule as shown in the equation

HCl	+	H_2O	$\longrightarrow$	H_3O^+	+	Cl^-
Hydrochloric acid	plus	water	yields	hydronium ion	plus	chloride ion

Note: The hydronium ion always has a +1 charge. The part of the acid that remains after the hydrogen leaves has a –1 charge. In the example above, the Cl^- is this leftover.

The negative ion of the acid can be a single element, as the Cl^- in the example above, or it can be a polyatomic ion as shown in the equation below. The process is the same in either case.

HNO_3	+	H_2O	$\longrightarrow$	H_3O^+	+	NO_3^-
Nitric acid	plus	water	$\longrightarrow$	hydronium ion	plus	nitrate ion

Note: The order in which the reactants or the products are written does not matter so long as all reactants are on the left and all products are on the right.

If an acid's formula includes two or more hydrogen atoms, write its reaction with water by allowing just one hydrogen atom to leave:

H_2SO_4	+	H_2O	$\longrightarrow$	H_3O^+	+	HSO_4^-
Sulfuric acid (battery acid)	plus	water	$\longrightarrow$	hydronium ion	plus	hydrogen sulfate ion

(More advanced texts deal with the other hydrogen atoms.)

EXERCISES

Complete the following equations. (Remember that it does not matter which product is written first.)

1. $HF + H_2O \longrightarrow$
2. $HClO_4 + H_2O \longrightarrow$
3. $H_3PO_4 + H_2O \longrightarrow$
4. $H_2SO_3 + H_2O \longrightarrow$
5. What product is always formed when water reacts with an acid?

Answers

1. $HF + H_2O \longrightarrow H_3O^+ + F^-$
2. $HClO_4 + H_2O \longrightarrow H_3O^+ + ClO_4^-$
3. $H_3PO_4 + H_2O \longrightarrow H_3O^+ + H_2PO_4^-$
4. $H_2SO_3 + H_2O \longrightarrow H_3O^+ + HSO_3^-$
5. a hydronium ion

BASES

Bases appear in our lives as soaps, oven cleaners, household cleansers, ammonia, and antacids, to name a few. The very word *antacid* conveys the idea that bases are somehow the opposite of acids. In fact, just as acids are defined as being H^+ donors, bases are defined as being H^+ acceptors. By removing a H^+ ion from neutral water (H_2O), bases convert water to the OH^- ion, called the *hydroxide ion*. Because bases have a bitter taste and a slippery feel caused by the hydroxide ion, it is safe to say that the makers of Hydrox cookies chose the name for some other reason.

To see how H^+ acceptance works, examine this reaction:

$$NH_3 + H_2O \longrightarrow NH_4^+ + OH^-$$

Ammonia plus water ⟶ ammonium ion plus hydroxide ion

Try to visualize a hydrogen ion leaving the water molecule and attaching itself to the ammonia (NH_3). The OH^- is what remains of the water molecule.

***Bases* are defined by their production of hydroxide (OH^-) ions when in water.**

EXERCISES

As a recap before we go on, answer these questions.

1. What ion is present in all water solutions of an acid? (Give the name and the formula.)
2. What ion is present in all water solutions of a base? (Give the name and the formula.)

Answers

1. the hydronium ion (H_3O^+)
2. the hydroxide ion (OH^-)

Example

Here is an Example of the reaction of a base with water. The basic nature of Tums and Rolaids is caused by the presence of calcium carbonate ($CaCO_3$). It reacts with water according to the equation

$$CaCO_3 + H_2O \longrightarrow Ca^{2+} + HCO_3^- + OH^-$$

Calcium carbonate + water ⟶ calcium ion + bicarbonate ion + hydroxide

The hydroxide ion produced by the reaction with water identifies the calcium carbonate as a base.

Hint: Make it a practice to look at every equation with an eye to figuring out what has happened. For example, in the equation above, ask yourself the following questions.

1. What happened to the water molecule? It lost a hydrogen ion.
2. What happened to the calcium ion (Ca^{2+})? The Ca^{2+} ion , originally embedded in a crystal, went into solution as a result of the reaction of its partner, carbonate ion (CO_3^{2-}) with the water.

3. What happened to the carbonate ion (CO_3^{2-})? It accepted a hydrogen ion.

4. Which species is the base? The carbonate ion (CO_3^{2-}).

Answers

Note: It's a good idea to keep Tables 2.1 and 2.3 handy.

EXERCISES

Each of these compounds or ions is a base. This means they can accept a hydrogen ion from the water molecule; so let this happen as you write the products.

1. $HCO_3^- + H_2O \longrightarrow$
2. $ClO^- + H_2O \longrightarrow$
3. $NH_2CH_3 + H_2O \longrightarrow$
4. The popular antacid Maalox is named for the combination of three ions in its formula. The first of these ions is magnesium. What are the other two ions?

Answers

1. $HCO_3^- + H_2O \longrightarrow H_2CO_3 + OH^-$
2. $ClO^- + H_2O \longrightarrow HClO + OH^-$
3. $NH_2CH_3 + H_2O \longrightarrow NH_3CH_3^+ + OH^-$
4. The other two ions in the Maalox name are Al for aluminum and ox for the hydroxide.

Other Bases

In some bases, you can see the hydroxide ion in the formula itself. For example, sodium hydroxide (NaOH) is the primary compound in Drano, a drain cleaner. When it is mixed with water, hydroxide ions are dispersed into the solution as the NaOH dissolves.

$$NaOH + H_2O \longrightarrow Na^+ + OH^- + H_2O$$

Sodium + water ⟶ sodium ion hydroxide ion water

These reactions are easier to write, but they do not show the acceptance of the hydrogen ion by the base. The presence of the hydroxide ion on the product side tells you that the sodium hydroxide is a base. It appears that the water just helps the hydroxide compound fall apart into its ions.

If there is more than one OH^- (hydroxide ion) in the beginning compound, the reaction looks like this:

$$Ba(OH)_2 + H_2O \longrightarrow Ba^{2+} + 2OH^- + H_2O$$

Barium hydroxide + water ⟶ barium + 2 hydroxide ions + water

Notice that the only real difference is the 2 placed before the OH^-.

Remember to include the ionic charges on the product side. By now you probably know that the hydroxide ion carries a –1 charge, and that allows you to determine the charge on the cation. However, you can also use Table 2.1.

EXERCISES

Complete each of the following equations for the reaction of bases with OH^- in the formula.

1. $KOH + H_2O \longrightarrow$
2. $Mg(OH)_2 + H_2O \longrightarrow$
3. $Al(OH)_3 + H_2O \longrightarrow$

Answers

1. $KOH + H_2O \longrightarrow K^+ + OH^- + H_2O$
2. $Mg(OH)_2 + H_2O \longrightarrow Mg^{2+} + 2OH^- + H_2O$
3. $Al(OH)_3 + H_2O \longrightarrow Al^{3+} + 3OH^- + H_2O$

EXERCISES

Combining the Ideas of Acid and Base

A. Examine the following equilibrium reaction (one that goes both ways) and answer these questions.

$$NH_3 \quad + \quad H_2O \longrightarrow NH_4^+ \quad + \quad OH^-$$

1. An acid is a hydrogen ion donor. Which of the molecules on the left loses (donates) a H^+ ion in moving from left to right?
2. What does the NH_3 do that makes it a base?
3. Notice the forward and reverse arrows in the equation. When NH_4^+ becomes NH_3, does it function as an acid or as a base? Why?
4. In the equation above label each of the four participants *acid* or *base* based on what you have learned.

B. Now, using the skills you have just taught yourself, look at the following equation and identify the two acids and the two bases.

$$HA \quad + \quad H_2O \longrightarrow H_3O^+ \quad + \quad A$$

Answers

A.

1. water
2. accepts a hydrogen ion
3. an acid because it donates a hydrogen ion
4. NH_3 + H_2O $\longrightarrow$ NH_4^+ OH^-

   Base acid $\longrightarrow$ acid base

B. HA + H_2O $\longrightarrow$ H_3O^+ A^-

acid base $\longrightarrow$ acid base

You may have noticed that water can act either as an acid or as a base, depending on what kind of compound it is with. **A substance that can act as an acid or as a base is said to be amphoteric.** In evaluating the role of water in a reaction, you have to look at the products of the reaction in order to decide if water functions as an acid or as a base.

THE DISSOCIATION OF WATER

For the most part, water molecules stay glued together in units of two hydrogen atoms and one oxygen atom. However, a very small number of water molecules dissociate. That is, they come undone and reassemble, like this:

$$2H_2O \longrightarrow H_3O \quad + \quad OH^-$$

This balanced equation indicates that a hydrogen ion (H^+) came from one water molecule, leaving one H and one O behind as OH^-. The loose H^+ attached itself to another water molecule to make H_3O^+, the hydronium ion (Figure 11.3).

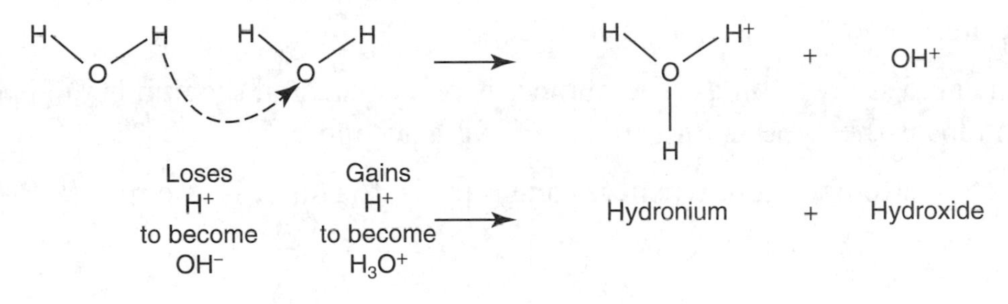

Figure 11.3: Hydronium ion formation

The small number of water molecules that exhibit this dissociation behavior is reflected in the size of the equilibrium constant for the equation above.

$$K_w = \frac{[H_3O^+][OH^-]}{[H_2O]^2} = 1 \times 10^{-14}$$

We do not use the molarity of the water because it is a pure substance, so

$$K_w = [H_3O^+][OH^-] = 1 \times 10^{-14}$$

Note: The very small size of 1×10^{-14} says that only a very few water molecules become H_3O^+ and OH^-, whereas the rest stay together as H_2O.

EXERCISE

Answer these questions about the dissociation of water.

1. When water dissociates, why is the equilibrium constant (K_w) so small?
2. In any amount of water, what can you say about the number of hydronium ions and the number of hydroxide ions?

3. What can you say about the ratio of hydronium ions to hydroxide ions in a bottle of water?

Answers

1. The equilibrium constant is small because almost all water molecules remain water molecules and do not become hydronium and hydroxide ions. You could also say that it is small because only a very small number of water molecules dissociate.
2. You could say two things: The number is very small, and the number of hydronium ions is the same as the number of hydroxide ions.
3. It is a one-to-one ratio. Whenever one is made, the other is also made.

UNEQUAL NUMBERS OF HYDRONIUM AND HYDROXIDE IONS

The number of hydronium ions (H_3O^+) and hydroxide ions (OH^-) are equal only when the solution is neither acidic nor basic. The following list is something you could have prepared yourself, knowing that hydronium ions are characteristic of acids and water and that hydroxide ions are characteristic of bases and water.

Neutral solution: $[H_3O^+] = [OH^-]$

Acidic solution: $[H_3O^+] > [OH^-]$

Basic solution: $[H_3O^+] < [OH^-]$

The Relationship of $[H_3O^+]$ to $[OH^-]$

Now that you know that $[H_3O^+][OH^-] = 1 \times 10^{-14}$, if you know the molarity of one of the ions, you can calculate the molarity of the other one. This is the algebraic equivalent of saying that if you know any two of the variables in the equation $a = bc$, you can solve for the third.

Examples

1. What is the molarity of the hydroxide ion if the molarity of the hydronium ion is 1×10^{-6}?

$[H_3O^+]\ [OH^-] = 1 \times 10^{-14}$

$[1 \times 10^{-6}]\ [OH^-] = 1 \times 10^{-14}$

$[OH^-] = \dfrac{1\times10^{-14}}{1\times10^{-6}}$ (Remember to subtract exponents when dividing)

$= 1 \times 10^{-8}$

2. Find the hydronium ion concentration if you know the hydroxide ion concentration (molarity) to be 2×10^{-5}.

$[H_3O^+][OH^-] = 1 \times 10^{-14}$

$[H_3O^+][2 \times 10^{-5}] = 1 \times 10^{-14}$

$[H_3O^+] = \dfrac{1\times10^{-14}}{2\times10^{-5}} = 5\times10^{-10}$

If you did not get the correct answer to this problem on your calculator, consult the calculator section in Unit 15.

EXERCISES

Answer the following questions about hydronium and hydroxide ions.

1. In all water solutions of acidic substances, is the hydronium ion or the hydroxide ion present in the larger amount?
2. If the hydronium ion concentration is 1×10^{-2}, what is the hydroxide ion concentration?
3. Are there basic ions in an acidic solution?
4. If the hydroxide ion concentration is 2.5×10^{-4}, what is the hydronium ion concentration?
5. What is the $[H_3O^+]$ in a solution of sodium hydroxide (NaOH) in which the $[OH^-]$ is 0.025 M? Is this solution acidic or basic?
6. What is the hydroxide ion concentration in a solution whose hydronium ion concentration is 4.5×10^{-5} M? Is this solution acidic or basic?

Answers

1. There are more hydronium ions in an acidic solution.
2. $[H_3O^+][OH^-]=1\times10^{-14}$

$$[OH^-]=\frac{1\times10^{-14}}{1\times10^{-2}}$$
$$=1\times10^{-2}$$

3. There are basic ions in an acidic solution, but there are more acidic ions than basic ions in an acidic solution. Remember that the result of multiplying the hydroxide concentration by the hydronium concentration is always 1×10^{-14}; so as one increases, the other decreases.
4. $[H_3O^+][OH^-]=1\times10^{-14}$

$$[H_3O^+]=\frac{1\times10^{-14}}{2.5\times10^{-4}}$$
$$=4\times10^{-11}$$

5. $[H_3O^+][OH^-]=1\times10^{-14}$

   $[H_3O^+][0.025]=1\times10^{-14}$

   $[H_3O^+]=4\times10^{-13}$

   The solution is basic because $[OH^-] > [H_3O^+]$.
6. $[H_3O^+][OH^-]=1\times10^{-14}$

   $[4.5\times10^{-3}][OH^-]=1\times10^{-14}$

   $[OH^-]=2.22\times10^{-12}$

This solution is acidic because there are more hydronium ions than hydroxide ions.

STRONG AND WEAK ACIDS AND BASES

Strong and *weak* are words used to describe acids and bases. Although you might think that strong means that a substance is more dangerous or more concentrated than a weak one, some weak acids are actually more hazardous than some strong acids. The terms *strong* and *weak* are involved in the concept of equilibrium. *Strong* means that the reaction of an acid or a base goes to completion. *Weak* means that the reaction of an acid or a base is an equilibrium situation.

EXERCISES

A. Think back over the skills you acquired in the unit on equilibrium and earlier in this unit. Answer the following questions about the reaction of a strong acid, HCl, with water.

1. Write a balanced equation for the reaction.
2. Should the reaction arrow be a one-way arrow or forward and reverse arrows? Why?
3. After a short time, will there be any HCl left over as an intact molecules?

B. From what you now know about strong acids and bases, you can figure out what weak acids and bases must be like.

1. Write an equation for the reaction of a weak acid, HX, with water.
2. After equilibrium has been established, are there any molecules of HX left over?
3. How can you tell if there is more HX or more H_3O^+ at equilibrium?
4. Which dissociates more: strong or weak acids and bases?

Answers

A.

1. $HCl + H_2O \longrightarrow H_3O^+ + Cl^-$
2. One way, because strong acids "go to completion," meaning that there is no reverse reactions.
3. All the HCl becomes products, and so there will be no HCl left.

B.

1. $HX + H_2O \longrightarrow H_3O^+ + X^-$
2. There are molecules of HX at equilibrium because the reverse reaction takes place, constantly replenishing the HX and the H_2O.
3. The size of the equilibrium constant indicates whether there are more products or more reactants.
4. Strong acids and bases dissociate 100 percent.

The Measuring Scale for Acids: pH

Acidity is measured by the concentration of hydronium ions using a measure called *pH*. We have become accustomed to seeing references to pH in advertising, product labeling, and swimming pool maintenance, for example. Shampoos are said to be pH-balanced, which certainly sounds like something that we would want. It means that the pH of the shampoo is the same as the pH of your scalp. As you teach yourself about pH in this section, keep an eye out for the mention of pH as you shop and watch television.

The pH scale is like a ruler marked from 0 to 14. This number span reflects the values of the equilibrium constant for the dissociation of water mentioned earlier in this unit. Recall the following equation.

$$K_w = [H_3O^+][OH^-] = 1 \times 10^{-14}$$

Each whole number on the pH scale matches a $[H_3O^+]$, as shown in Table 11.1.

TABLE 11.1: RELATIONSHIP OF $[H_3O^+]$ TO pH

pH	$[H_3O^+]$		
1	1×10^{-1}	=	0.1
2	1×10^{-2}	=	0.01
3	1×10^{-3}	=	0.001
4	1×10^{-4}	=	0.0001
5	1×10^{-5}	=	0.00001
6	1×10^{-6}	=	0.000001

The pattern shown in the table continues to a pH of 14.

EXERCISES

Answer the following questions dealing with pH.

1. If the pH is 8, what is the $[H_3O^+]$?
2. If the $[H_3O^+]$ is 1×10^{-7}, what is the pH?
3. If the hydronium ion molarity is 1×10^{-2}, what is the pH?
4. Which is more acidic, a pH of 1 or a pH of 3?
5. Every time the pH goes up by 1 (for example, pH of 4 to pH of 5), what happens to the amount of hydronium ions?
6. How many times more acidic is a pH of 3 than a pH of 5?
7. In a solution whose pH is 8, are there more hydronium ions or more hydroxide ions?
8. If you add water to a solution whose pH is 3, will the pH go up or down?

Answers

1. 1×10^{-8} or 0.00000001
2. 7
3. 2
4. 1 (because 0.1 M is greater than 0.001 M)
5. You have a tenth as many.
6. 100 (10×10)
7. More hydroxide ions. (The solution is basic.)
8. The pH will go up, less acidic

For an illustration of where everyday household items fit on the pH scale, see Figure 11.4.

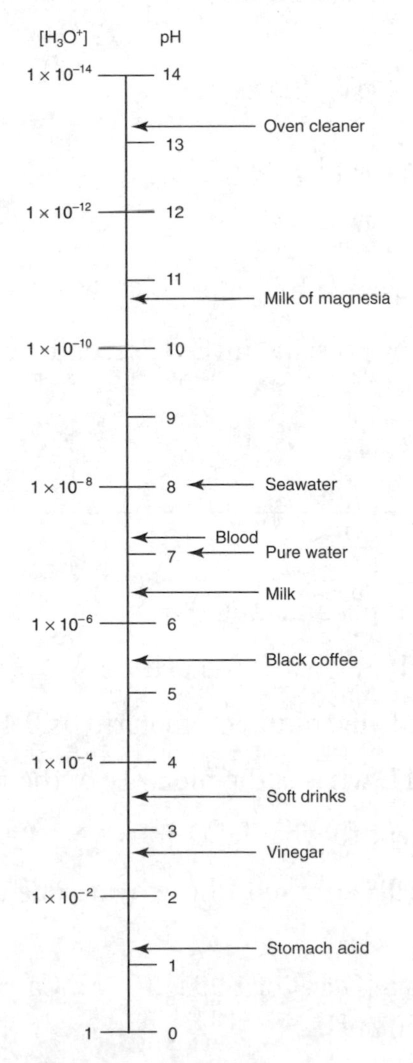

Figure 11.4: pH scale

You may have already noticed that in the discussion of pH so far every hydronium ion concentration was a 1×10 to some power, such as 1×10^{-3}. What happens if the concentration is some other number, such as 2.5×10^{-2} (0.025)? This goes to the definition of pH, which says that **pH is the negative logarithm of the molarity of the hydronium ion.** In equation format it looks like this:

$$pH = -\log[H_3O^+]$$

Problems can be of two varieties: Find the pH when you know the $[H_3O^+]$, and find the $[H_3O^+]$ when you know the pH.

To calculate the pH from the $[H_3O^+]$:

- Enter the $[H_3O^+]$.*
- Press the log key.
- Press the +/– key to change the sign.

*If the $[OH^-]$ is given, calculate the $[H_3O^+]$.

To calculate the $[H_3O^+]$ from the pH:

- Enter the pH.
- Press the +/– key to change the sign.
- Take the inverse log by pressing inverse, second, or shift and then log.

EXERCISES

Answer these questions about pH and $[H_3O^+]$.

1. If the $[H_3O^+]$ is 2.19×10^{-4}, what is the pH?
2. Calculate the pH if the hydronium ion molarity is 9.18×10^{-11}.
3. If a solution's pH is 4.11, what is the molarity of the hydronium ion?
4. A pH of 9.54 is equivalent to what $[H_3O^+]$?
5. What is the pH of a NaOH solution whose hydroxide concentration is 4.22×10^{-3} M?
6. A mystery solution has a $[OH^-]$ of 1.99×10^{-7} M. Calculate both its hydronium ion concentration and its pH.
7. A solution of nitric acid (HNO_3) has a $[H_3O^+]$ of 2.19×10^{-4}. What is its pH?

8. A solution of potassium hydroxide (KOH) has a hydroxide ion concentration of 0.02 M. What is its pH?

9. If the pH of blood is 7.3, calculate the $[H_3O^+]$.

10. The pH of stomach acid is 2.15. Calculate both its hydronium ion and its hydroxide ion concentrations.

Answers

1. pH = 3.66
2. pH = 10.04
3. $[H_3O^+] = 7.76 \times 10^{-5}$
4. $[H_3O^+] = 2.88 \times 10^{-10}$
5. Because you need to know the hydronium ion concentration in order to calculate the pH, the first step is

$$[H_3O^+][4.22 \times 10^{-3}] = 1 \times 10^{-14}$$
$$[H_3O^+] = 2.37 \times 10^{-12}$$

Then

$$pH = -\log(2.37 \times 10^{-12})$$
$$= 11.63$$

6.

$$[H_3O^+][1.99 \times 10^{-7}] = 1 \times 10^{-14}$$
$$[H_3O^+] = 5.03 \times 10^{-8}$$
$$pH = -\log(5.03 \times 10^{-8})$$
$$= 7.3$$

7.

$$pH = -\log(2.19 \times 10^{-4})$$
$$= 3.66$$

8.

$$[H_3O^+][0.02] = 1 \times 10^{-14}$$
$$[H_3O^+] = 5 \times 10^{-13}$$
$$pH = -\log(5 \times 10^{-13})$$
$$= 12.30$$

9.

$[H_3O^+]$ = – inverse log of pH
= 7.3, +/–, shift, log
= 5.01×10^{-8}

10.

pH = 2.15
$[H_3O^+]$ = 2.15, +/–, shift, log
= 7.08×10^{-3}
$[OH^-]$ = $(1 \times 10^{-14})/(7.08 \times 10^{-3})$
= 1.41×10^{-12}

BUFFERS

A buffer solution is one that changes pH only slightly when small amounts of a strong acid or a strong base are added. A buffer solution is created by either

- A weak acid plus its salt (such as CH_3COOH and NaCOOH) *or*
- A weak base plus its salt (such as NH_3 and NH_4Cl)

The buffer solution is able to control the pH so that pH stability is maintained. For example, when 0.01 mole of HCl is added to 1 liter of pure water, the pH changes from 7 to 2. This is a change of 5 pH units. However, when the same amount of HCl is added to a buffered solution of CH_3COOH (acetic acid) and NaCOOH (sodium acetate), the pH changes by less than 0.1 pH unit.

The control of pH is vital in many applications. The pH of human blood must be within the narrow range of 7.35 to 7.45 to ensure survival. Acidosis, a decrease in the pH of the blood, can be brought on by diabetes, heart failure, and kidney failure. Alkalosis, an increase in the pH of the blood, is symptomatic of altitude sickness, hyperventilation, and vomiting. A blood pH outside the normal values presents a serious health crisis.

NEUTRALIZATION

Remember that a neutral solution is neither acidic nor basic. In the process of neutralization, an acid and a base are combined and the resulting solution is neutral. A neutralization we all remember is the use of baking soda (a base) on a bee sting (formic acid). Once neutralized, the acid no longer burns.

In neutralization, an acid plus a base produces water plus a salt.

Notice that this statement says "a salt," not "salt." The reason that this is important is that *salt* is commonly used to mean NaCl, whereas the term *a salt* can mean any of a large number of compounds having cations and anions. A salt can be $MgSO_4$, KF, $CaCl_2$, and so on.

You have likely heard the term *neutralization* in advertisements, such as "neutralizes 47 times its weight in excess stomach acidity" in reference to a leading antacid (base). Neutralization is the strategy behind applying lime (a base) to soil to combat too much acidity.

Predicting the Products of a Neutralization Reaction

Follow these steps in order to create a balanced equation for the combination of an acid and a base.

1. Write the first product as water (even if it doesn't seem right).
2. Write the second product (the salt) as a combination of the other two ions.
3. Use the charges to figure out the subscripts of the salt.
4. Balance the equation.

Example

$HNO_3 + Ba(OH)_2 \longrightarrow H_2O + Ba^{2+} NO_3^-$

$HNO_3 + Ba(OH)_2 \longrightarrow H_2O + Ba(NO_3)_2$

$2HNO_3 + Ba(OH)_2 \longrightarrow 2H_2O + Ba(NO_3)_2$

Tip: Save the coefficient that goes in front of the water for last.

EXERCISES

For each of the following neutralization reactions, write the products and then balance the equation.

1. $HBr + NaOH \longrightarrow$
2. $H_2SO_4 + NaOH \longrightarrow$
3. $HCl + Mg(OH)_2 \longrightarrow$
4. $HNO_3 + Al(OH)_3 \longrightarrow$
5. $H_3PO_4 + KOH \longrightarrow$

Answers

1. $HBr + NaOH \longrightarrow H_2O + NaBr$
2. $H_2SO_4 + 2NaOH \longrightarrow 2H_2O + Na_2SO_4$
3. $2HCl + Mg(OH)_2 \longrightarrow 2H_2O + MgCl_2$

4. $3HNO_3 + Al(OH)_3 \longrightarrow 3H_2O + Al(NO_3)_3$

5. $H_3PO_4 + 3KOH \longrightarrow 3H_2O + K_3PO_4$

As you reflect on this discussion of acids and bases, recall that it was all about the hydrogen ion. Acids donate hydrogen ions, which are really just protons. Bases accept hydrogen ions, and so acids and bases are different sides of the same coin.

UNIT 12

Equilibrium Reactions

Equilibria and Le Chatelier's Principle

In a chemical reaction at *equilibrium*, reactants become products at the same time that products are turned back into reactants. In other words, the reaction goes both forward and backward.

The idea of equilibrium hinges on the concept of reaction rates. In chemistry rate refers to how much something changes in a unit of time. The "something" that changes is the concentration of a reactant or a product, usually expressed as molarity. The unit of time is generally the second, although any unit of time can be used. Sometimes it is desirable to manipulate the rate of a reaction in order to speed it up or slow it down. The factors that affect the rate of a reaction are temperature, concentration, surface area, and the use of a catalyst.

RATE FACTORS

Increasing the temperature generally speeds up a reaction. Lowering the temperature of food by refrigeration causes a decrease in the rate of the bacterial action responsible for food spoilage. The factors that affect reaction rates are not desirable or undesirable in themselves and can be useful manipulators to serve a desired purpose.

Increasing the concentration of a reactant results in an increased rate for that chemical reaction. This is largely an "opportunity" phenomenon. For example, if reactant A and reactant B combine in a reaction, the more often they can "find each other," the faster the reaction proceeds. Increasing the concentration of either reactant increases the speed at which the product is formed. Similarly, you can decrease the rate of a reaction by reducing the concentration of one or more of the reactants.

The idea of surface area can be illustrated by steel wool. If you carefully separate a pad of steel wool into individual strands and pieces, it will cover a surprisingly large area. However, if you have the same weight of steel in a chunk, it will cover a much smaller area. Steel wool has a larger surface area than a chunk of steel. This large surface area is responsible for increasing the rate of reaction in its combustion with oxygen. Steel wool can be ignited with a match, but a chunk of steel cannot be ignited and can only be warmed with a match.

By definition, a *catalyst* is a substance that influences the rate of a reaction while remaining unchanged itself. For example, the human body contains thousands of catalysts, called enzymes, without which chemical reactions in the body would take place too slowly to sustain life. A catalyst works because it provides a new pathway for a reaction. This pathway has a lower activation energy than does the original route taken by reactants in becoming products, as shown in Figure 12.1.

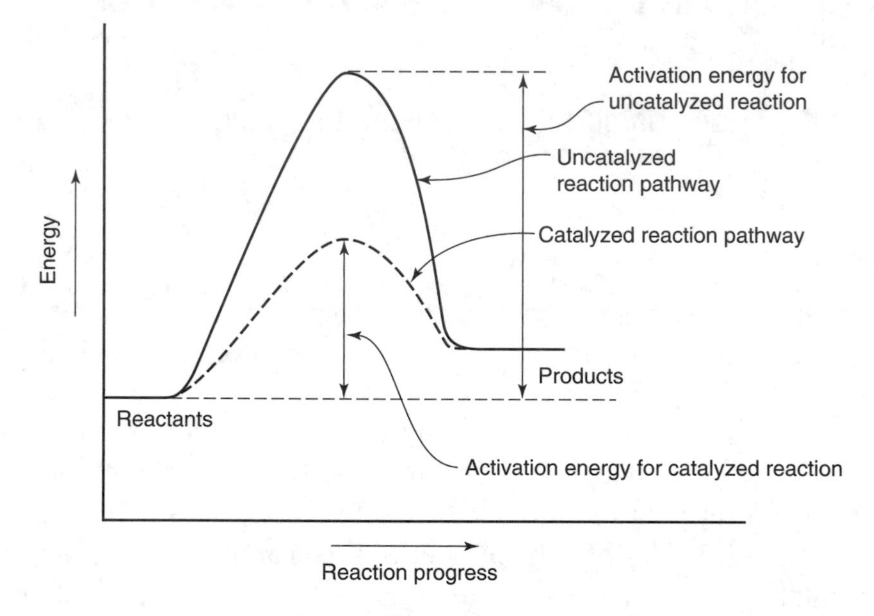

Figure 12.1: The catalyst pathway

EXERCISES

A. Examine Figure 12.1 in order to answer these questions.

1. In this case, which has more energy: the products or the reactants?
2. Does the uncatalyzed reaction have an activation energy?
3. What does a catalyst do with respect to the activation energy for a reaction?
4. What does the distance between the peaks of the two curves represent?

B. Answer the following questions about reaction rate factors.

1. Hydrogen peroxide is commonly used in homes to disinfect cuts and scrapes. It decomposes over time and therefore loses its potency. Which of the rate factors mentioned above can be used to slow its decomposition reaction?
2. Experienced campers start a fire by using tinder, which is very small, shredded pieces of wood, rather than logs. Which of the rate factors plays a part in this strategy?
3. The chemical reaction that creates rust has water as one of its reactants. Old cars abandoned in the desert rust very slowly compared with cars in other climates. Which of the factors mentioned above explains this?
4. Automobiles have a device called a catalytic converter. Its purpose is to promote a reaction by which noxious exhaust gases are changed to less environmentally harmful ones. What does this catalyst do to the activation energy?

Answers

A.

1. The products have more energy, as they appear higher on the y (energy) axis.
2. Both reactions have an activation energy, which is the distance between the reactant energy and the top of the curve in both cases.
3. A catalyst lowers the activation energy for a reaction.
4. The distance between the two curve peaks represents the difference between the activation energies for the catalyzed and the uncatalyzed reactions.

B.

1. The temperature rate factor can be used to decrease the decomposition reaction rate by storing the hydrogen peroxide in a refrigerator.
2. Tinder has a much larger surface area than the surface areas of logs, thereby improving the reaction rate.
3. Water is a reactant in the reaction that produces rust. In a desert climate the amount of water is seriously reduced. A reduction in the concentration of a reactant causes a reduction in the rate of the reaction.
4. The catalyst in catalytic converters reduces the activation energy of the reaction.

JUST WHAT IS EQUAL IN AN EQUILIBRIUM?

Imagine a school with 25 people in the building. Twenty people are inside a classroom, and the other 5 people are out in the hall. No one can enter or leave the building, but every minute 2 of the people in the hall come into the classroom and 2 of the people in the classroom go out into the hall. This creates an equilibrium. The rate, that is, the number of people entering the classroom per minute, is equal to the number of people leaving the classroom per minute. Although the number of people in each location does not change, the identity of these people changes within the bounds of the 25 people in the school building. The numbers used in this example do not have any special significance. For instance, an equilibrium would exist if every minute 5 people went in and 5 went out of the room. Similarly, an equilibrium would exist if there were always 7 people in the hall and 18 people in the room. What makes this an equilibrium system is that an equal number of people go into the room and out of the room in a given time period.

The same idea applies to a chemical reaction *at equilibrium*. While reactants are becoming products, products are returning to reactants. The reaction in which reactants become products is called the *forward reaction,* and the reaction in which products become reactants is called the *reverse reaction*. At **equilibrium, the rate of the forward reaction is equal to the rate of the reverse reaction.** When a reaction is an equilibrium reaction, there are two arrows between the reactant side of the equation and the product side of the equation. One arrow points to the products, and one arrow points to the reactants. All chemical equilibrium reactions are written with a two-way arrow. For example:

$$A + B \rightleftharpoons C + D$$

In order for an equilibrium to be reached, a system must be closed. In a *closed system,* new reactants cannot be added and products cannot be removed. In the school building example above, the system is closed because other people cannot enter the building and no one can leave the building. The only permitted movement is within the building. If other people could come into the building or some could leave the building, an equilibrium could not exist.

An equilibrium is said to be *dynamic,* which means that the forward and reverse reactions constantly take place at the same rate. In the earlier example involving people moving around in a school building, their constant movement creates a dynamic situation. However, the numbers of persons in each location is static, or unchanging.

EXERCISE

Suppose that there is an equilibrium reaction R + S $\rightleftharpoons$ T + U. Mark each of the following statements true or false.

1. The rate at which R and S become T and U is equal to the rate at which T and U become R and S.
2. Once T and U form, they do not combine to make R and S.
3. At equilibrium some of R, S, T, and U is present.

Answers

1. true
2. false
3. true

REACTIONS THAT ARE NOT EQUILIBRIA

Not all reactions are equilibria reactions. Such reactions are said to go to completion, meaning that once the products are made, the reaction is complete, or finished. When you put ingredients (reactants) together to make a cake, stir them up, and bake them for the required time, the reaction goes to completion. The reactants do not reappear in their original form of eggs, flour, and so on. All the atoms you began the process with are now part of the cake, which is the product.

Some chemical reactions cannot be in equilibrium because of their own physical limitations. For example, a reaction in an open container in which one or more of the participants is a gas cannot be in equilibrium because the gas can escape, preventing the system from being closed. Another impediment to the establishment of an equilibrium is the presence of a precipitate as one of the products. Because the precipitate is by its very nature virtually insoluble, it is not available for participation in the reverse reaction.

EQUILIBRIUM EXPRESSIONS

An equilibrium expression is a mathematical expression that describes the relative amounts of products with respect to the relative amounts of reactants.

For a reaction like this:

$$aA \quad + \quad bB \quad \rightleftharpoons \quad cC \quad + \quad dD$$

where the lowercase letters are the balanced-equation coefficients and the uppercase letters represent the chemical formulas for the reactants and products, the equilibrium expression is written as

$$K_{eq} = \frac{[A]^a[B]^b}{[C]^c[D]^d}$$

The brackets indicate molarity, so [A] means moles of A per liter of solution. (See Unit 7 for a review of molarity if necessary.) K_{eq} is called the equilibrium constant.

Once equilibrium has been reached, if there is a higher concentration of products than of reactants, the equilibrium constant has a value greater than 1 because the numerator is larger than the denominator.

A higher equilibrium concentration of reactants compared to products creates an equilibrium constant with a value less than 1 because the numerator is small compared to the denominator. Many chemical reactions that take place in our bodies are reactions in which a healthy situation is marked by such an equilibrium. The size of the equilibrium constant is a function of the reaction itself, with each reaction having its own K_{eq} at a given temperature.

Example

The following equilibrium reaction is given:

$$N_2 \quad + 3H_2 \quad \rightleftharpoons \quad 2NH_3$$

Write the equilibrium expression:

$$K_{eq} = \frac{[NH_3]^2}{[N_2][H_2]^3}$$

Given that the K_{eq} for this particular reaction is 3.5×10^8, is there a larger concentration of products or of reactants? (Equilibrium constants are generally written in scientific notation.)

There is a larger concentration of products because the equilibrium constant is larger than 1, meaning that the numerator is larger than the denominator.

EXERCISES

A. Answer the following questions for the reaction

$$H_2 \quad + \quad 2I \quad \rightleftharpoons \quad 2HI$$

1. What is the equilibrium expression?

2. How can you tell whether this reaction is an equilibrium reaction?

3. What is equal in this reaction?

4. The K_{eq} for this reaction is 5.41×10^1. Does this favor the products or the reactants?

B. Write the equilibrium expression for each of the following reactions.

1. $N_2 \quad + \quad 3Cl_2 \quad \leftrightarrows \quad 2NCl_3$

2. $2SO_2 \quad + \quad O_2 \quad \leftrightarrows \quad 2SO_3$

Answers

A.

1. $\dfrac{[HI]^2}{[H_2][I]^2}$

2. An equilibrium is shown by the use of two-way arrows: $\rightleftharpoons$.

3. The rate at which HI is being produced is equal to the rate at which H_2 and the I are being made by the reverse reaction.

4. Because the equilibrium constant is a number greater than 1, the products are favored over the reactants.

B.

1. $K_{eq} = \dfrac{[NC1_3]^2}{[N_2][CI_2]^3}$

2. $K_{eq} = \dfrac{[SO_3]^2}{[SO_2]^2[O_2]}$

EQUILIBRIUM PROBLEMS

The solution of equilibrium problems is based on algebraic manipulation of the equilibrium constant expression. In typical equilibrium problems the unknown is one part of the K_{eq} equation shown above and the other parts are given.

Examples

1. For the reaction

$$A_2 + 2B \rightleftharpoons 2AB$$

the equilibrium concentrations are $[A_2] = 0.0045$, $[B] = 0.95$, and $[AB] = 1.5$. Calculate the value of the equilibrium constant, K_{eq}.

Step 1. Write the equilibrium expression:

$$K_{eq} = \frac{[AB]^2}{[A_2][B]^2}$$

Step 2. Substitute the given values where appropriate:

$$K_{eq} = \frac{(1.5)^2}{(0.0045)(0.95)^2}$$

Step 3. Solve for the unknown:

$$K_{eq} = 5.54 \times 10^2$$

2. The equilibrium constant is 8.96×10^{-2} for the reaction

$$PCl_5 \rightleftharpoons PCl_3 + Cl_2$$

The equilibrium concentration of PCl_5 is 3.5×10^{-3} M, and the equilibrium concentration of PCl_3 is 0.25 M. Calculate the equilibrium concentration of Cl_2.

Step 1. Write the equilibrium expression:

$$K_{eq} = \frac{[PCl_3][Cl_2]}{[PCl_5]}$$

Step 2. Substitute the given values:

$$8.96 \times 10^{-2} = \frac{(0.25)[Cl_2]}{3.5 \times 10^{-3}}$$

Step 3. Solve for the unknown:

$$(8.96 \times 10^{-2})(3.5 \times 10^{-3}) = 0.25[Cl_2]$$

$$0.00125 \text{ M} = [Cl_2]$$

3. When the equilibrium constant is given for a particular reaction, the equilibrium constant for the reverse reaction is the reciprocal of the equilibrium constant for the forward reaction.

The equilibrium constant for the following reaction is 5.0×10^1.

$$H_2 + I_2 \rightleftharpoons 2HI$$

What is the equilibrium constant for the reaction

$$2HI \rightleftharpoons H_2 + I_2$$

Because this reaction is the reverse of the given reaction, the equilibrium constant is the reciprocal, or $(1/5.0) \times 10^1 = 2.0 \times 10^{-2}$.

EXERCISES

Answer the following questions dealing with the equilibrium constant.

1. Calculate the value of the equilibrium constant for the reaction below with given equilibrium concentrations of $[N_2] = 0.04$ M, $[O_2] = 0.008$ M, and $[NO] = 4.7 \times 10^{-4}$ M.

$$N_2 + O_2 \rightleftharpoons 2NO$$

2. Calculate the molarity of HF in the reaction below if the value of the equilibrium constant is 2.1×10^3 and the equilibrium concentrations of H_2 and F_2 are both 0.0018 M.

$$H_2 + F_2 \rightleftharpoons 2HF$$

3. The equilibrium constant for the reaction below is 6.6×10^{-1}.

$$N_2O_4 \rightleftharpoons 2NO_2$$

What is the equilibrium constant for the reaction

$$2NO_2 \rightleftharpoons N_2O_5$$

Answers

1. $$K_{eq} = \frac{[NO]^2}{[NO_2][O_2]}$$
$$= \frac{(4.7\times10^{-4})^2}{(0.04)(0.008)}$$
$$= 6.90\times10^{-4}$$

2. $$K_{eq} = \frac{[HF]^2}{[H_2][F_2]}$$
$$2.1\times10^3 = \frac{[HF]^2}{(0.0018)(0.0018)}$$
$$6.80\times10^{-3} = [HF]$$

3. If the equilibrium constant for the forward reaction is 6.6×10^{-1}, the equilibrium constant for the reverse reaction is the reciprocal, $(1/6.6) \times 10^{-1} = 1.51$.

LE CHATELIER'S PRINCIPLE

In 1888 the French chemist Henri Le Chatelier studied equilibrium systems and came up with an idea for describing what happens to an equilibrium when it is stressed by a change. There are times when it is important to take a qualitative look at a chemical reaction to see how it responds to changes.

Although what he concluded was quite lengthy, the essence of it, called **Le Chatelier's Principle, is that if a stress is applied to a system that is at equilibrium, the system will shift to reach a new equilibrium that partially offsets the impact of the stress**. *Stresses* are changes that can be applied to chemical reactions and include changes in temperature, changes in pressure, and changes in concentrations of reactants or products. The word *shift* refers to a change in equilibrium. A shift to the right causes more products to be formed, whereas a shift to the left causes the formation of more reactants.

Stresses can be used to manipulate an equilibrium reaction in order to create a shift that will result in the formation of either more products or more reactants. Shifting an equilibrium to the right should be the plan if your goal is the manufacture of one or more of the product substances. Forming more reactants, or shifting the equilibrium to favor the reverse reaction could be the focus in the preparation of prescription drugs needed to regulate one of our body's chemical reactions.

Concentration as a Stress

If you change the concentration of a reactant by adding more of it, the equilibrium shifts to the right in order to "use up" some of the newly added reactant. If one of the products is removed from the system, the equilibrium shifts to the right in order to make up for the decrease in product.

Temperature as a Stress

Some reactions produce heat as a product, such as

$$A + B \rightleftharpoons C + D + \text{heat}$$

These reactions are called exothermic, meaning that they create heat. Removing heat by cooling such a reaction is a stress that would cause the equilibrium to shift to the right in an attempt to "replace" the lost heat.

Endothermic reactions require heat. A typical endothermic reaction is written

$$A + B + \text{heat} \rightleftharpoons C + D$$

The heat can be thought of as a reactant, so that heating such a reaction shifts the equilibrium to the right, favoring the products.

Pressure as a Stress

When the reactants and products are all gases, a change in pressure causes the equilibrium system to shift in the direction that relieves the stress. For example, in the reaction

$$N_2 + 3H_2 \rightleftharpoons 2NH_3$$

the reactant side of the equation has 4 moles (1 N_2 and 3 H_2), whereas the product side has only 2 moles. The pressure that a gas produces is a result of the number of moles present. If a stress is placed on this reaction in such a way as to increase the pressure, the equilibrium tries to relieve the stress by decreasing the pressure. The right side of the equation has fewer moles, so the shift is to the product side where fewer moles result in lower pressure. Pressure changes in a gaseous system are generally affected by a change in volume. When the volume is increased, the pressure decreases because the molecules have more room, hitting the sides of their container less often. If the volume is decreased, the pressure increases, just as squeezing a balloon causes the pressure inside to become greater.

EXERCISES

Answer the following questions dealing with Le Chatelier's principles.

1. A + heat $\rightleftharpoons$ B + C

a. If this reaction is cooled, in which direction does the equilibrium shift?

b. If some of B is taken out of the system, in which direction does the equilibrium shift?

c. If A, B, and C are all gases and the volume of the reaction is reduced, in which direction does the equilibrium shift?

2. Given: the following equilibrium reaction with all the participants as gases.

$4NH_3 + 5O_2 + \text{heat} \rightleftharpoons 4NO + 6H_2O$

For each of the stresses listed below, state whether the equilibrium shifts right, shifts left, or undergoes no change.

a. heating the system

b. adding NO

c. adding NH_3

d. removing water

e. increasing the volume

Answers

1. a. Left, because cooling the system decreases the amount of one of the reactants, heat, and so the equilibrium shifts to make more heat.

b. Right, because as one of the products (B) is removed, the equilibrium shifts to try to replace the lost product.

c. Left, because a reduction in volume creates an increase in pressure. In order to relieve this stress, the equilibrium shifts to the side with fewer moles to decrease the pressure.

2. a. Right, because heat is treated as a reactant. Adding more reactant shifts the equilibrium to the right.

b. Left, because adding more product causes the equilibrium to shift to the reactant side to relieve the stress caused by the presence of more product.

c. Right, because increasing the concentration of a reactant causes the equilibrium to shift in the direction that uses up the new reactant.

d. Right, because loss of the product, water, causes the equilibrium to attempt to replace the water.

e. Right, because increasing the volume decreases the pressure. In order to restore the pressure, the system shifts to the side that has more moles. The product side has 10 moles in comparison to 9 moles for the reactant side.

UNIT 13

Organic Chemistry

Organic Nomenclature and Formula Writing, Isomers, Consumer Connections, and Functional Groups

This unit deals with the essentials of organic chemistry. That means that the topics you will need for continuation of your chemical pursuits are presented here. Organic chemistry is the chemistry of carbon compounds. There are only a few carbon-containing compounds, such as carbon dioxide and carbon monoxide, that do not fall into the organic category. Gigantic corporations, such as Dow Chemical and DuPont, are largely producers of organic chemicals. In our everyday lives we are surrounded by organic substances such as antifreeze, alcohol, pharmaceuticals, lawn care products, cosmetics, nylon, Velcro, artificial sweeteners, and foods, to name just a few. In fact, we humans are mostly organic: muscles, tendons, skin, liver, brain, and so on. Chemists who shop the produce aisles of grocery stores must wonder about the "organic" section. All produce is organic, not just that grown without pesticides. The marketers of produce grown without pesticides have somehow taken the name *organic,* even though the true meaning of the word is much wider. There are presently several million organic compounds, with the number constantly growing.

The vastness of organic chemistry necessitates that there be a system of organization for the naming and grouping of compounds according to their functions. This system is called the International Union of Pure and Applied Chemistry *(IUPAC) system* and uses prefixes that indicate the number of carbon atoms present in a molecular formula (Table 13.1). Memorizing these prefixes will be helpful to you.

TABLE 13.1: CARBON GROUP PREFIXES

Prefix	Number of Carbon Atoms
meth-	one
eth-	two
prop-	three
but-	four
pent-	five
hex-	six
hept-	seven
oct-	eight

ALKANES

The simplest carbon compounds are called *alkanes*. **Alkanes are compounds of hydrogen and carbon and have only single bonds.** In naming these compounds, use the correct prefix and then add -ane (for alkane). For example, the alkane having three carbon atoms is propane. Now you know that the propane gas used in outdoor grills has molecules composed of three carbon atoms and some hydrogen atoms all held together by single bonds. The number of hydrogen atoms can be determined by looking up carbon in the periodic table and seeing that it has four valence electrons. In order to satisfy the octet rule (Unit 7), carbon needs to share four other electrons, therefore making four bonds.

The Lewis structures for the alkane having one carbon atom are as follows.

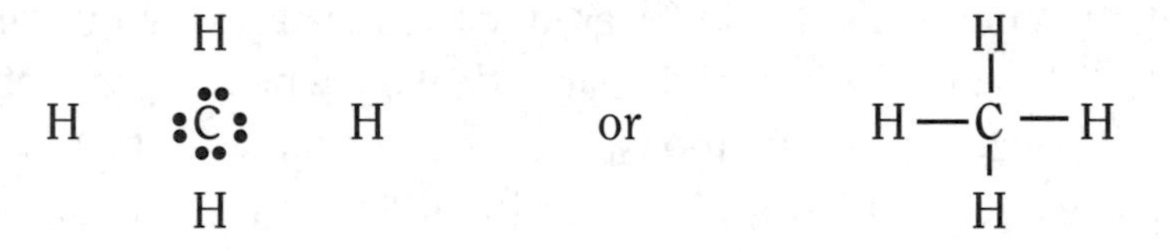

In general, each carbon atom has four bonding electrons, and so you would expect to show four lines (bonds) coming from each carbon. For example, propane is

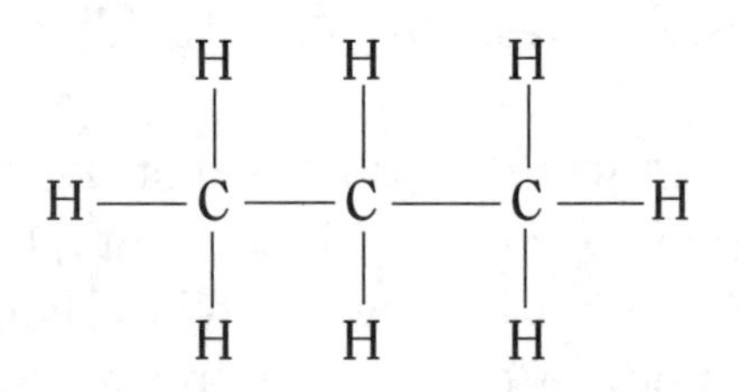

This is propane's *structural formula*. Propane can also be written as its molecular formula, C_3H_8. Assume for now that the carbon atoms are in a straight line.

EXERCISES

A. Provide the correct name for each of these formulas.

1. C_8H_{18}
2. C_2H_6
3. C_5H_{12}

B. Draw the structural formulas for each of the following.

1. hexane
2. butane
3. heptane
4. Write the molecular formulas for your answers to questions 1, 2, and 3 in part B.

Answers

A.

1. octane
2. ethane
3. pentane

B.

1.

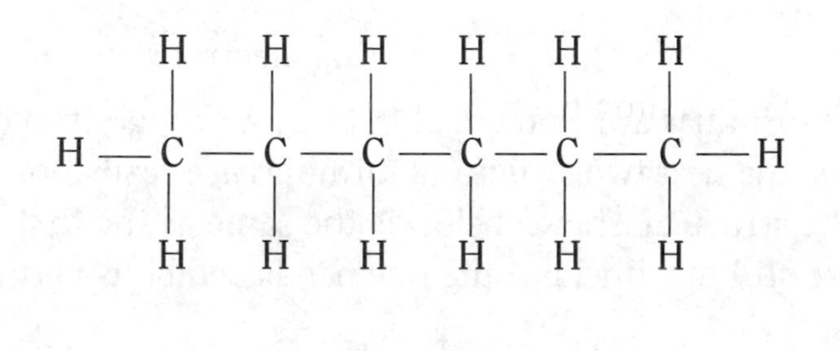

2.

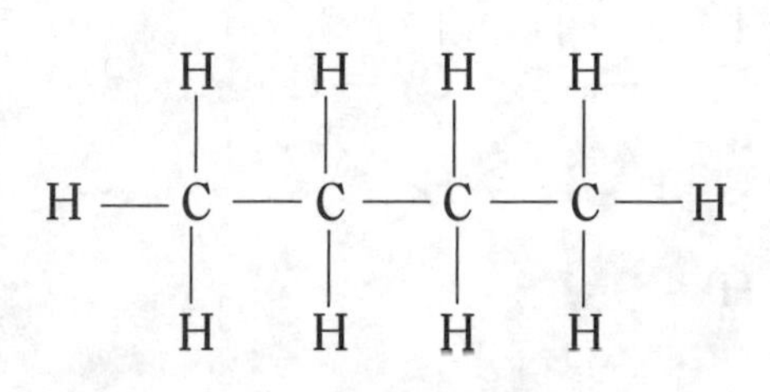

3.

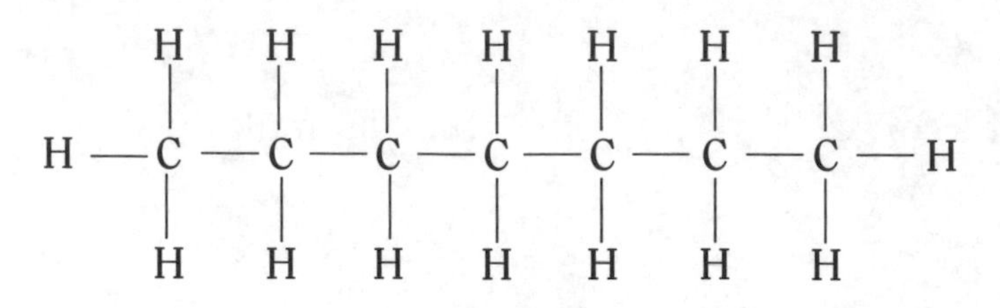

4. C_6H_{14}, C_4H_{10}, C_7H_{16}

ISOMERIC ALKANES

Isomers are compounds having the same molecular formulas but different structural formulas. To understand the idea of isomerism, imagine that you have a small pile of Lego blocks that can be put together in a variety of ways. The same principle is involved when considering an isomer of an organic compound. Take the same number of atoms of each element and put them together in a different manner, making certain that each atom has the correct number of bonds. For example, butane can be written in either of two ways:

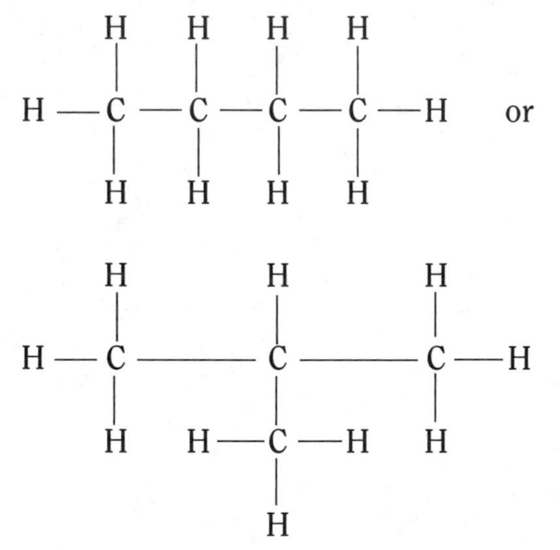

These two structural formulas are isomers of butane. Although it may seem that there are other possibilities, the next two structural formulas are really not different from the two shown above. The structure shown below is the same as the first one shown above. The carbon atoms are still in a line, but the line bends, which is permissible.

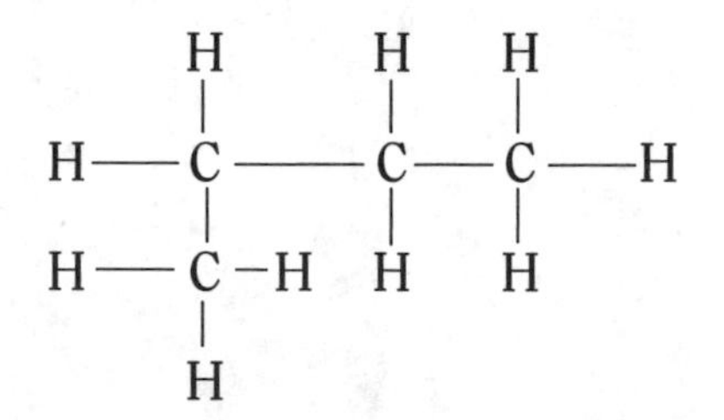

This structure is the same as the second one shown above, but it is viewed at 180 degrees.

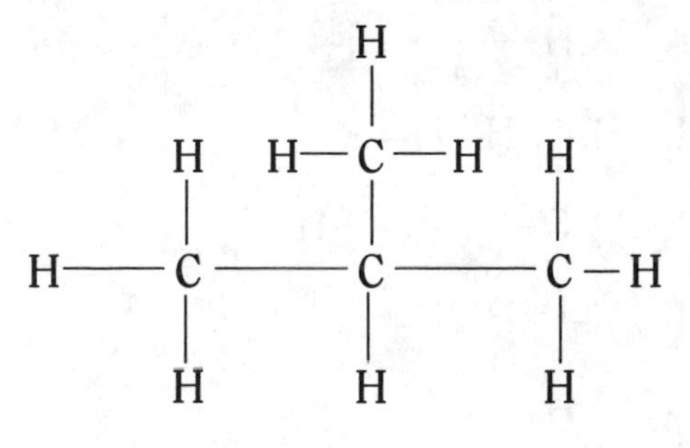

EXERCISES

Answer the following questions dealing with isomers.

1. For which alkanes can there be no isomers?
2. What are the possible isomers for pentane (C_5H_{12})?

Answers

1. Methane, ethane, and propane have no isomers because the small number of carbon atoms permits no other arrangements.
2. Pentane has three isomers:

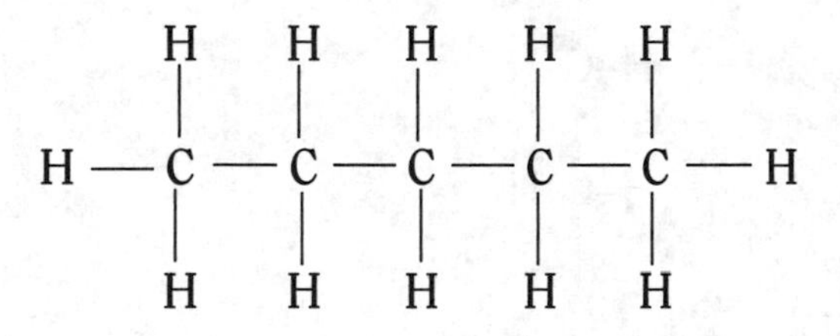

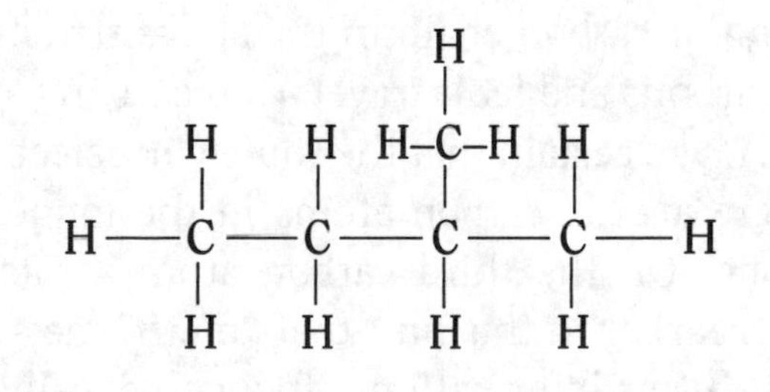

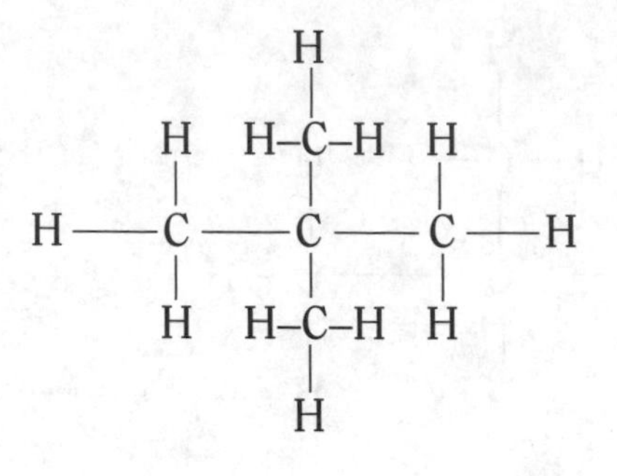

Naming Alkanes

In naming alkanes with carbon atoms in a straight line, use the correct prefix to reflect the number of carbons plus -ane, such as butane for the alkane having four carbon atoms. If the carbons are not in a straight line, find the longest continuous line of carbons. Number the carbon atoms in this line starting at the end that gives the lowest number for the location where something other than a hydrogen atom is attached.

Examples

1.

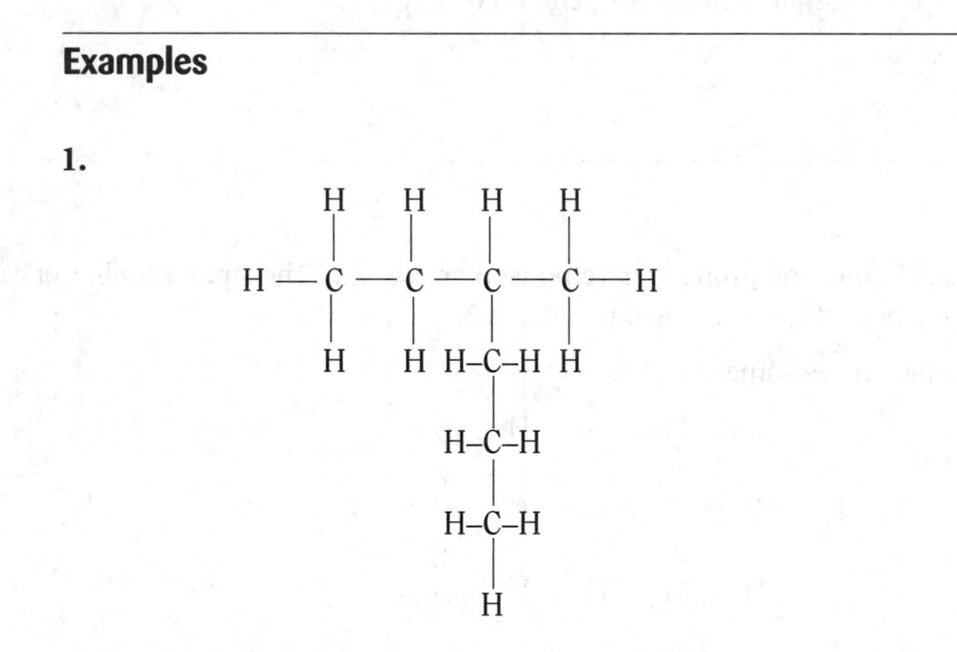

In looking for the longest carbon chain in the above structure, put your pencil on one of the carbons and let it travel from carbon to carbon without leaving the paper. Make certain that the route you select is the longest possible. Because there are six carbon atoms in the longest chain, this compound is a hexane. On the third carbon atom a methyl group is attached. (Recall from earlier in this unit that "meth" means one carbon atom.) In order to fit all this information into a name, call this molecule *3-methylhexane*. The *3* tells the reader where something is attached to the hexane chain, and the *methyl* indicates that this addition is a one-carbon group.

2.

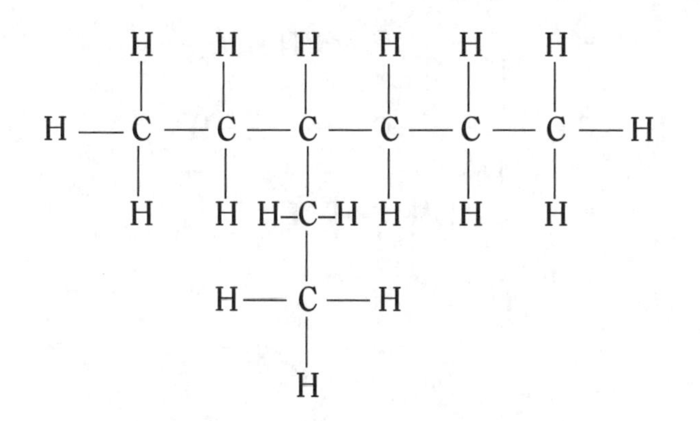

The name for this formula is 3-ethylhexane. The longest chain is the horizontal line of six carbon atoms (the hexane), and a two-carbon group (the ethyl) is attached at the third carbon of the main chain.

3.

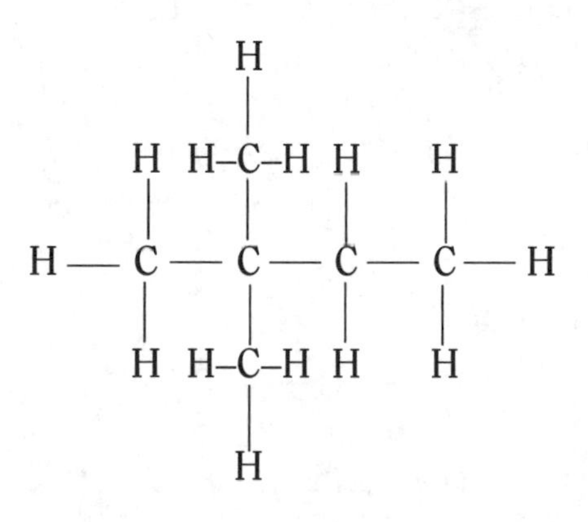

Some molecules have more than one group attached to the primary chain. In these cases, prefixes such as di- and tri- designate the number of such multiple groups. Examine the structure above to see why it is called 2,2-dimethylbutane.

EXERCISES

A. Name the following alkanes.

1.

```
          H
          |
    H   H-C-H   H
    |     |     |
H — C  —  C  —  C — H
    |     |     |
    H     H     H
```

2.

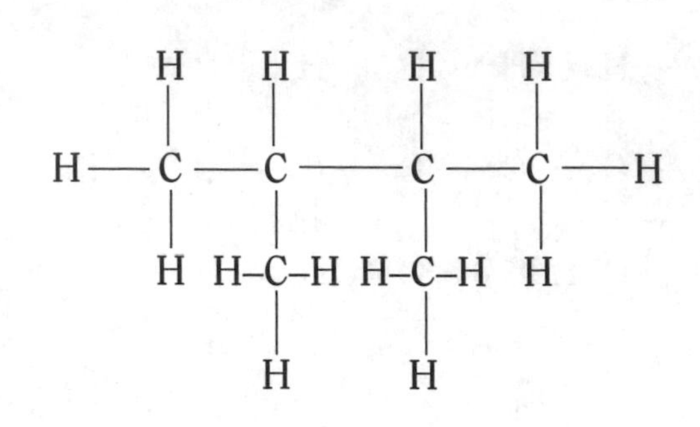

3.

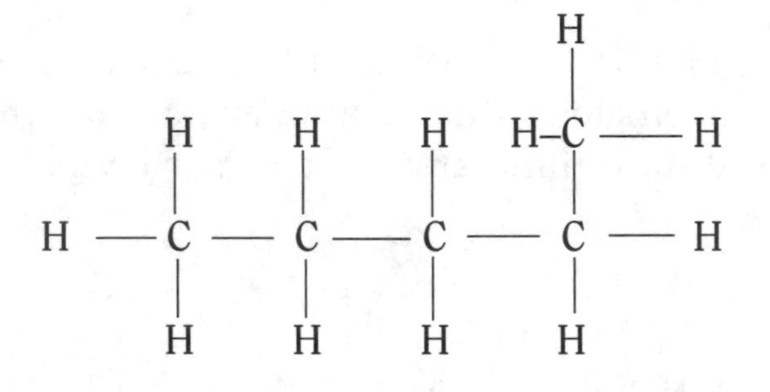

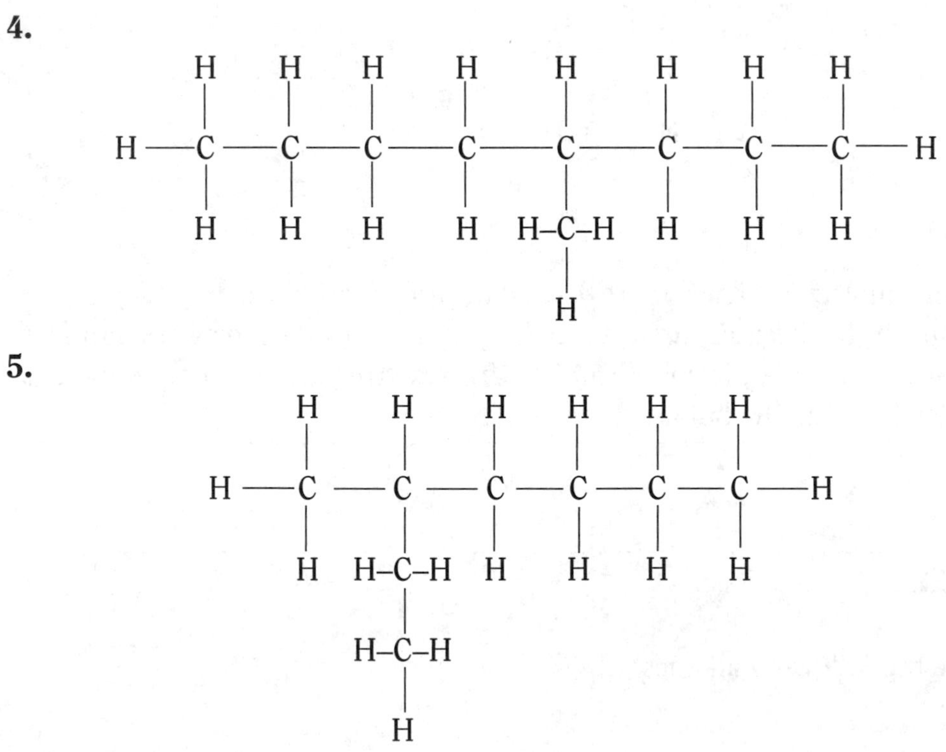

B. Draw the structural formulas for the following alkanes.

1. 3-ethylpentane

2. 3-methylhexane

3. 2-ethyloctane

4. 3,3-dimethylpentane
5. 2,4-dimethylheptane

Answers

A.

1. 2-methylpropane
2. 2,3-dimethylbutane
3. pentane
4. 4-methyloctane
5. 3-methylpentane

B.

1.

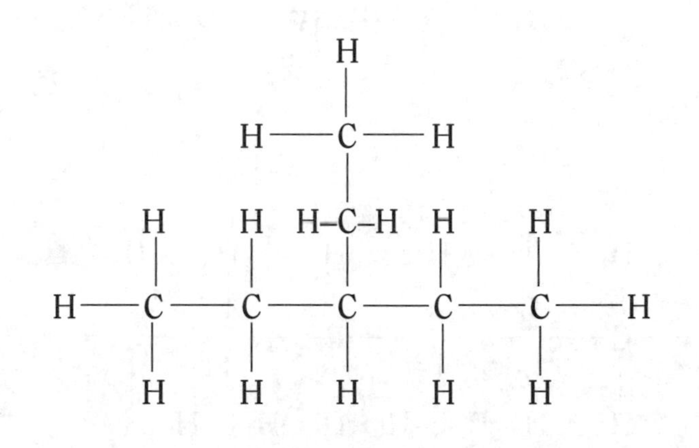

2.

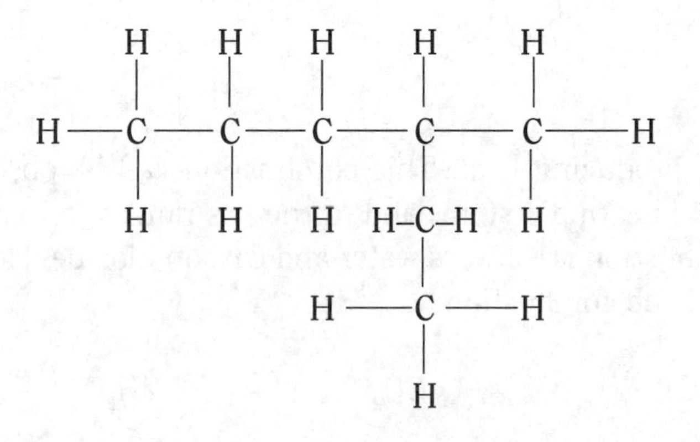

3.

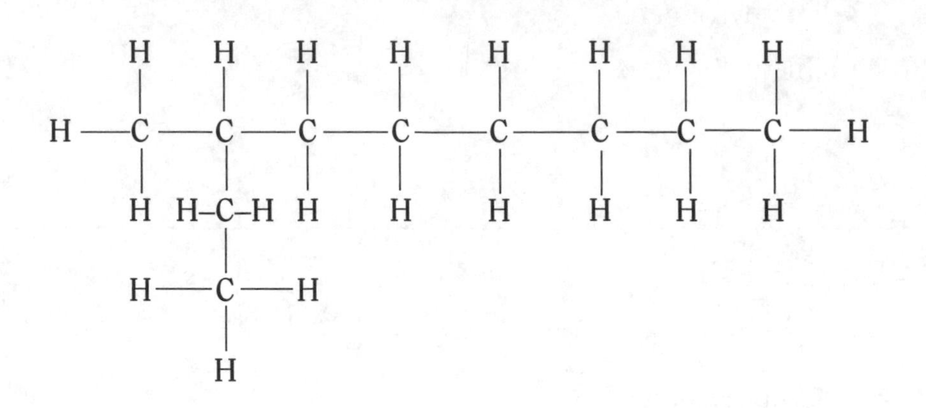

4.

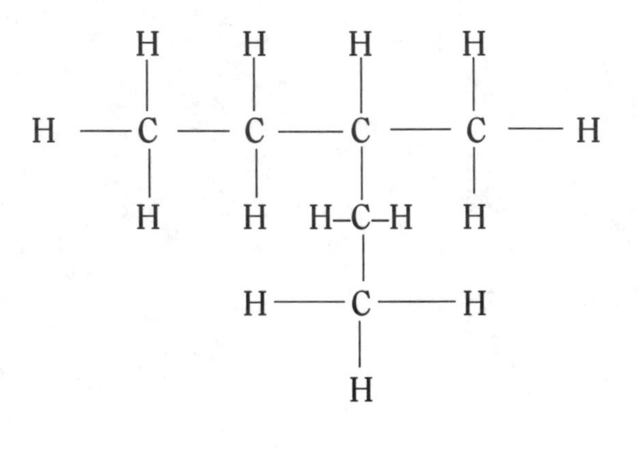

5.

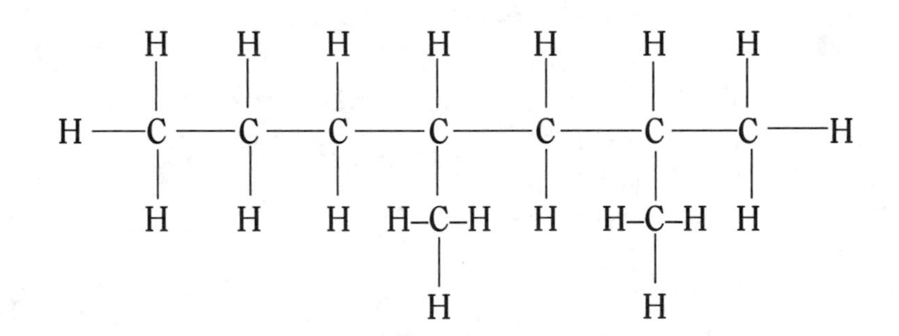

Alkane Combustion

Alkanes react with oxygen, producing heat. This combustion can be observed in the burning of methane in home heating systems and in the gas ranges in many kitchens. The products of alkane combustion are always water and carbon dioxide. Here is a sample balanced equation for alkane combustion.

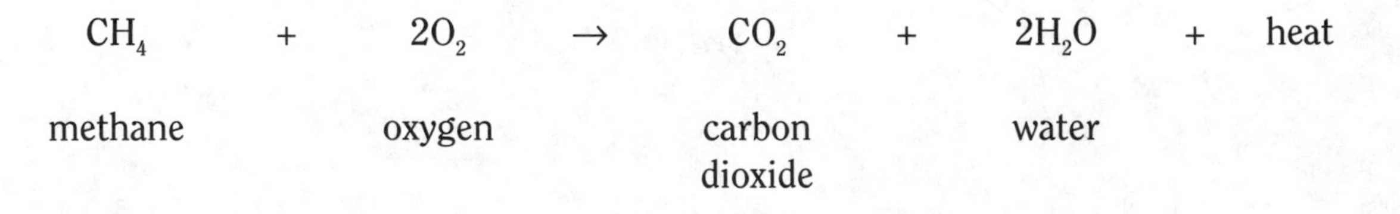

EXERCISE

Complete the following problems dealing with alkane combustion.

1. Write and balance the equation for the combustion of propane.
2. Write and balance the equation for the combustion of butane. (This reaction occurs when a butane lighter is lit.)

Answers

1. $C_3H_8 + 5O_2 \rightarrow 4H_2O + 3CO_2$
2. $2C_4H_{10} + 13O_2 \rightarrow 10H_2O + 8CO_2$

ALKENES

Alkenes are compounds of carbon and hydrogen having at least one double bond between carbon atoms. Their names utilize the prefixes you have already learned, with the ending changed to -ene. As you examine these examples, recall that each carbon requires four bonds and each hydrogen atom requires one bond.

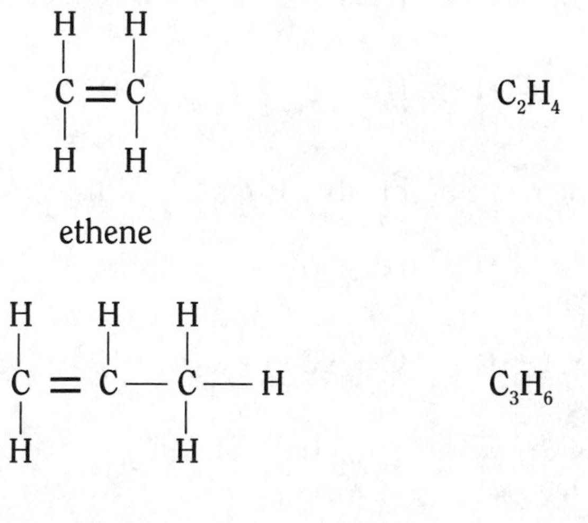

ethene

propene

In alkenes having four or more carbon atoms, there must be a way to indicate the location of the double bond. This is accomplished once again by numbering the carbon atoms in the longest chain. For example, 2-hexene is

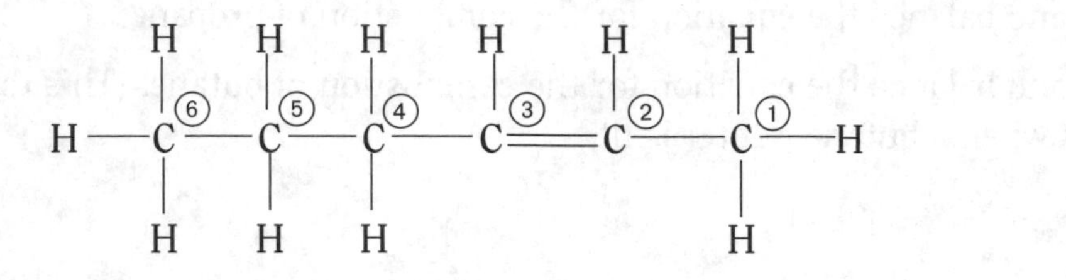

Remember to number from the end that gives the lower number for the location of the double bond.

EXERCISES

Answer the following questions about alkenes.

1. What is the name of this structure?

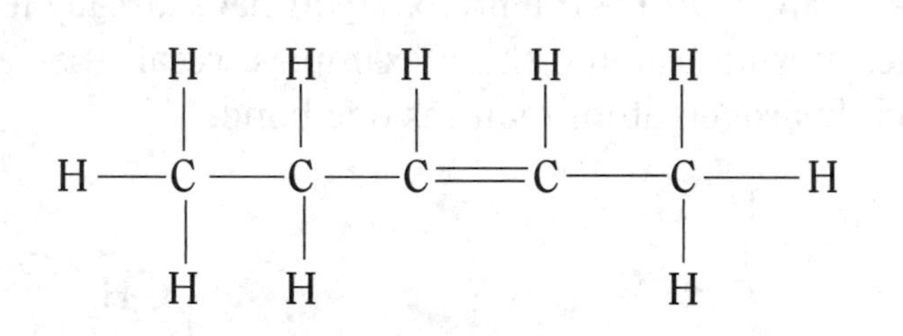

2. What is the name of this structure?

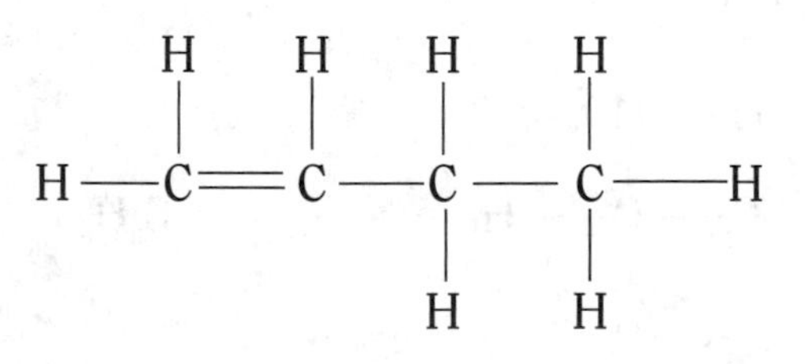

3. Draw the structural formula for 3-octene.

4. Draw the structural formula for 3-hexene.

5. Write the molecular formula for 3-hexene.

6. Could there be a 4-hexene?

Answers

1. 2-pentene
2. 1-butane
3.

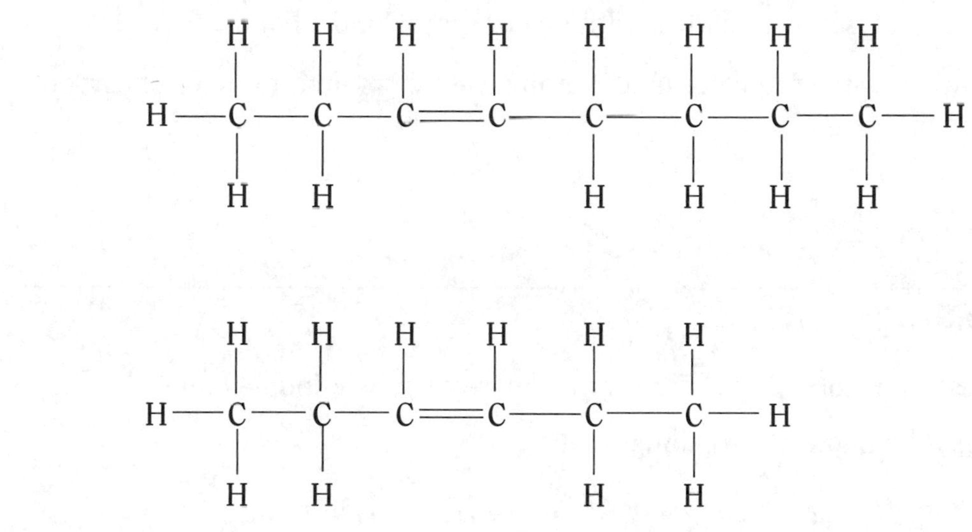

4.

5. The molecular formula is $C_6 H_{12}$.
6. No. Numbering from the end resulting in the smaller number would still make it 3-hexene.

Some alkenes have more than one double bond. When this happens, use numbers separated by commas to show the double-bond locations. Then use one of the prefixes before -ene to show the number of double bonds:

Prefix	Number of Double Bonds
di-	two
tri-	three

For example, the name for

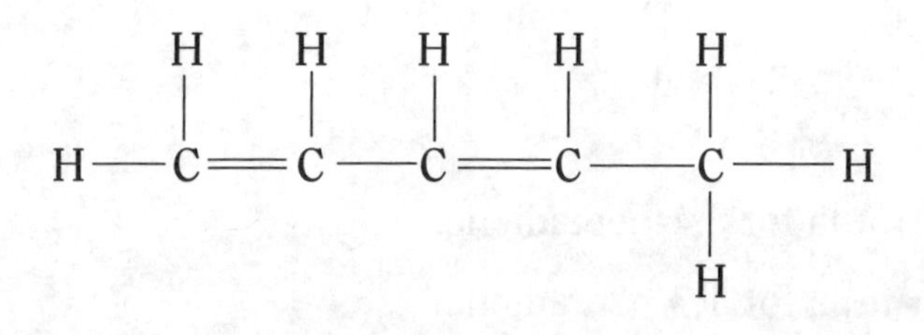

is 1,3-pentadiene:

- 1 and 3 indicate the carbon atoms where each double bond begins.
- penta indicates five carbon atoms.
- di indicates two double bonds.
- -ene indicates a hydrocarbon with one or more double bonds.
- Follow the same sequence in all naming (numbers first, number of carbons next, and so on).

EXERCISES

Answer these questions about alkenes with more than one double bond.

1. What is the name of this compound?

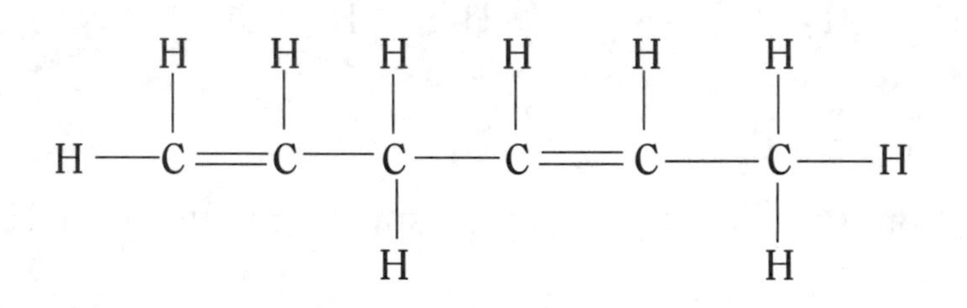

2. What is the name of this compound?

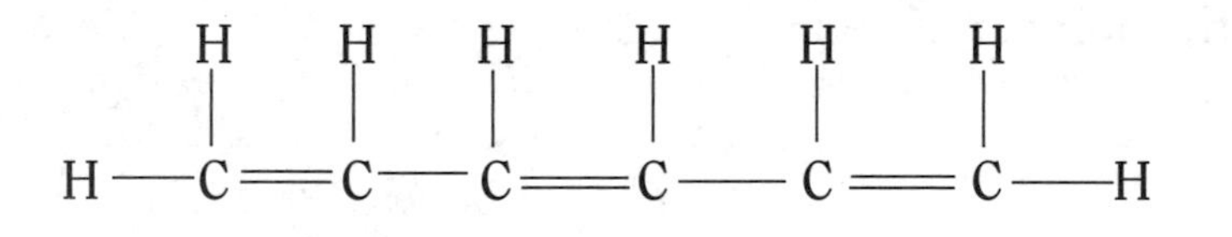

3. What is the name of this compound?

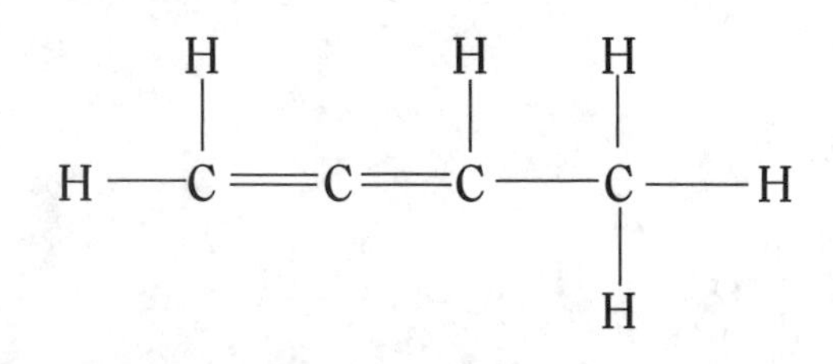

4. Draw the structural formula for 3,4-heptadiene.
5. Draw the structural formula for 1,3-pentadiene.
6. Draw the structural formula for 1,2,4-pentatriene.

Answers

1. 1,4-hexadiene
2. 1,3,5-hexatriene
3. 1,2-butadiene
4.

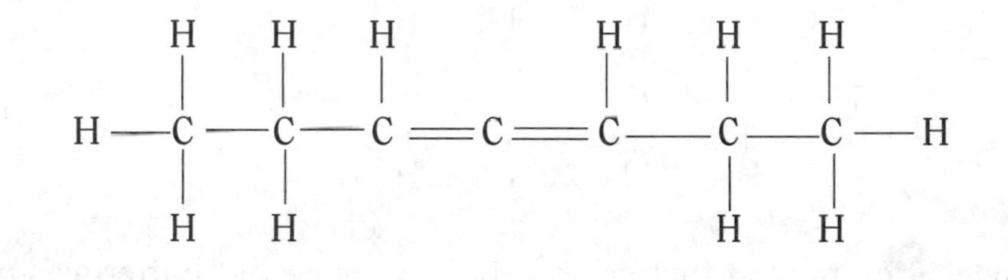

5.

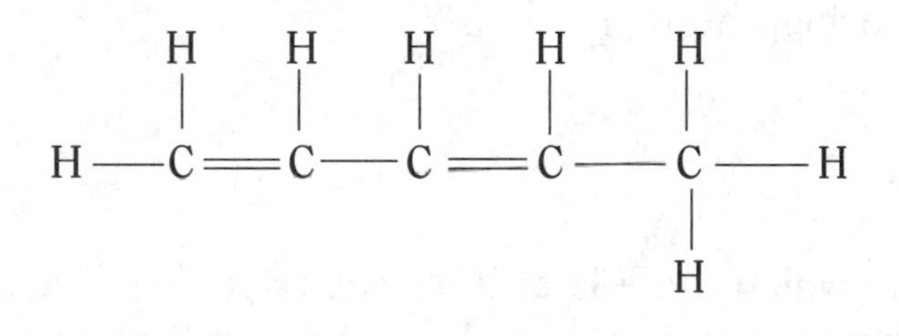

6.

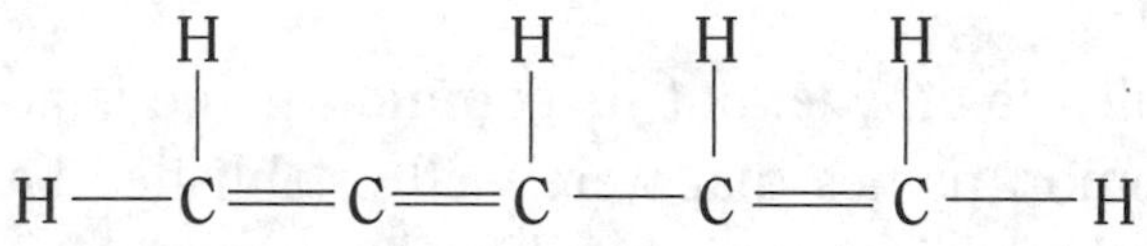

```
     H        H    H    H
     |        |    |    |
H — C == C == C — C == C — H
```

SATURATED AND UNSATURATED CHEMICAL BONDS

The terms *saturated* and *unsaturated* appear on virtually every nutrition label. They refer to the chemical bonds in fats such as butter, margarine, olive oil, and so on. In a saturated fat, the bonds are single bonds, whereas an unsaturated fat is characterized by the presence of one or more double bonds. In general, saturated fats are solid and unsaturated fats are liquids. Over the years there has been scientific controversy involving the health implications of each.

Hydrogenation is the process whereby hydrogen is added to a double bond, changing it to a single bond as in the following example.

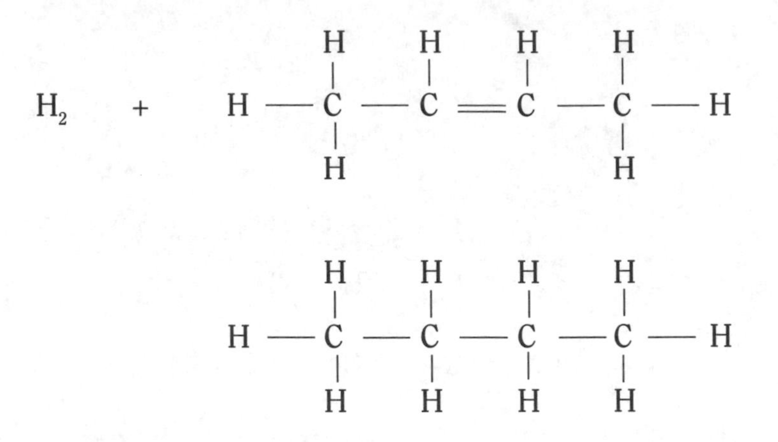

It is the hydrogenation process that creates the creamy peanut butter we enjoy today. The peanut butter of an earlier generation had a layer of peanut oil on top that had to be stirred into the peanut butter before spreading. (As an early radio commercial touted, "Better things for better living, through chemistry.")

ALKYNES

Alkynes are hydrocarbons in which there is at least one triple bond. Acetylene is an alkyne that you have likely heard of, as in an acetylene torch. Its formula is

$$\mathrm{H{-}C{\equiv}C{-}H}$$

and so its name would normally be ethyne, but its common name is acetylene. Some organic compounds have common names that were well established before the more formal naming system was developed. The triple bond in alkynes makes them extraordinarily reactive and hazardous. For this reason they are not a part of most familiar commercial products.

ALKYL GROUPS

The enormous variety of organic compound possibilities creates a need for simplifying the system for their description. If you remove one hydrogen atom from an alkane, you will have what is called an *alkyl group*. For example:

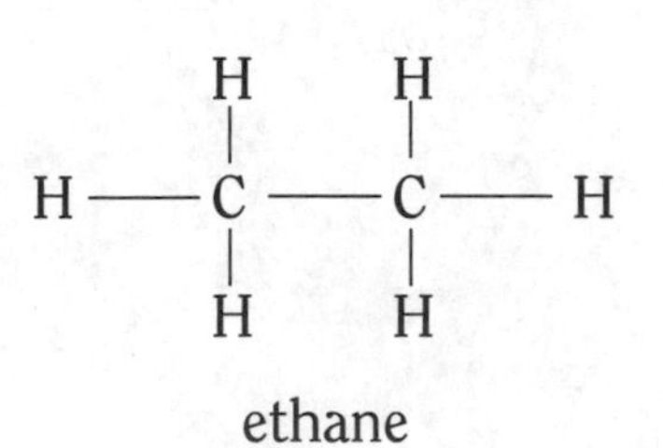

ethane

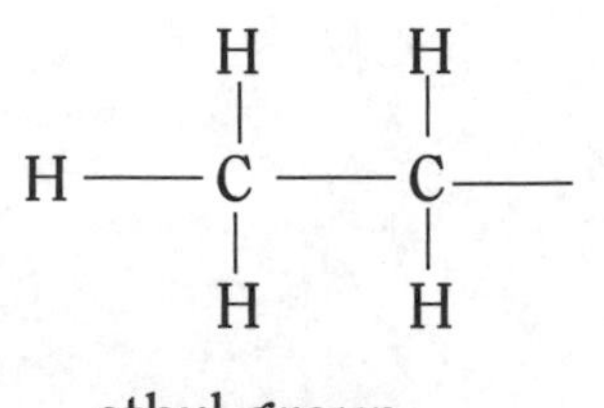

ethyl group

These alkyl groups are symbolized by the letter R. If you want to show that there are two R groups, use R and R' (called R prime). In other words, R can stand for any one of the alkyl groups: methyl, ethyl, propyl, and so on. As we continue discussing the idea of functional groups, you will see how R is used.

FUNCTIONAL GROUPS

Functional groups are groups of atoms whose presence causes molecules to have certain properties. Perhaps the best way to explain a functional group is to present an example. One functional group is the combination of one oxygen atom and one hydrogen atom (OH). When this group takes the place of one or more of the hydrogen atoms in an alkane, the result is an *alcohol.*

ALCOHOLS

In symbolic language, an alcohol is ROH, where R can be any alkyl group and OH is the defining functional group for alcohols. In writing the molecular formula for alcohols, the OH is written last. For example, the formula for methanol is CH_3OH rather than CH_4O. Keeping the functional group intact in formula writing alerts the reader that the compound is an alcohol.

Sometimes alcohols are present in our lives in unexpected places. Cholesterol is an alcohol whose structure is shown in Figure 13.1.

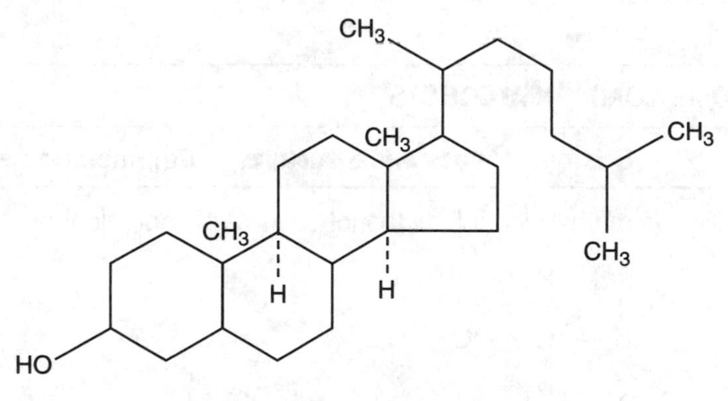

Figure 13.1: Cholesterol

Antifreeze is primarily ethylene glycol. Although it is an automotive necessity year-round, its extreme toxicity makes it one of the most hazardous chemicals found in our homes and garages. The structural formula for ethylene glycol is

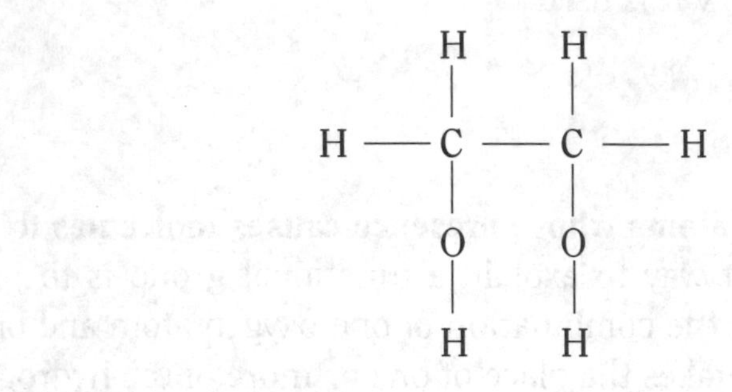

Naming Alcohols

The naming of alcohols is accomplished by

- Naming the parent alkane
- Removing the final -e
- Adding an -ol
- Placing a number in front of the name to indicate the location of the OH group. If the OH is at either end of the molecule, no number is needed.

Table 13.1 lists some common alcohols with their scientific names as well as their common names and their structures.

TABLE 13.1: SOME COMMON ALCOHOLS

Formula	Scientific Names and Structure	Common Name
CH_3OH	methyl alcohol, methanol H H — C — O — H H	wood alcohol
C_2H_5OH	ethyl alcohol, ethanol H H H — C — C — O — H H H	grain alcohol

Formula	Scientific Names and Structure	Common Name
C_3H_7OH	propyl alcohol, 2-propanol	rubbing alcohol

```
          H
          |
     H    O    H
     |    |    |
H -- C -- C -- C -- H
     |    |    |
     H    H    H
```

EXERCISES

A. Write the structural formula and molecular formula for each of the following alcohols.

1. 2-butanol

2. propanol

3. 2-octanol

4. butanol

5. 3-pentanol

B. Name the following alcohols

1.

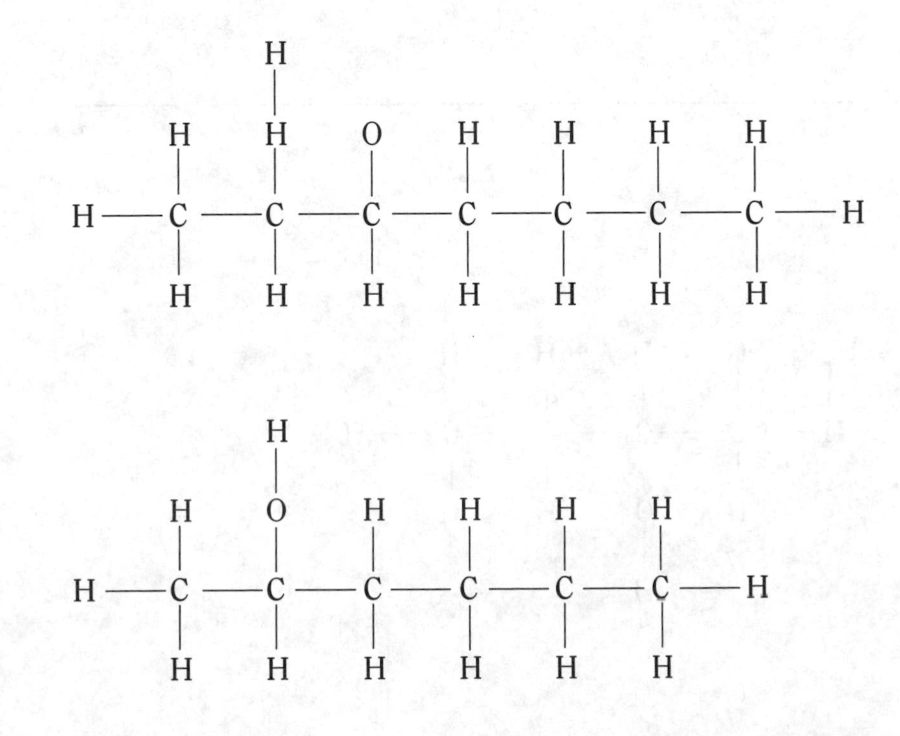

2.

3.

4.

5.

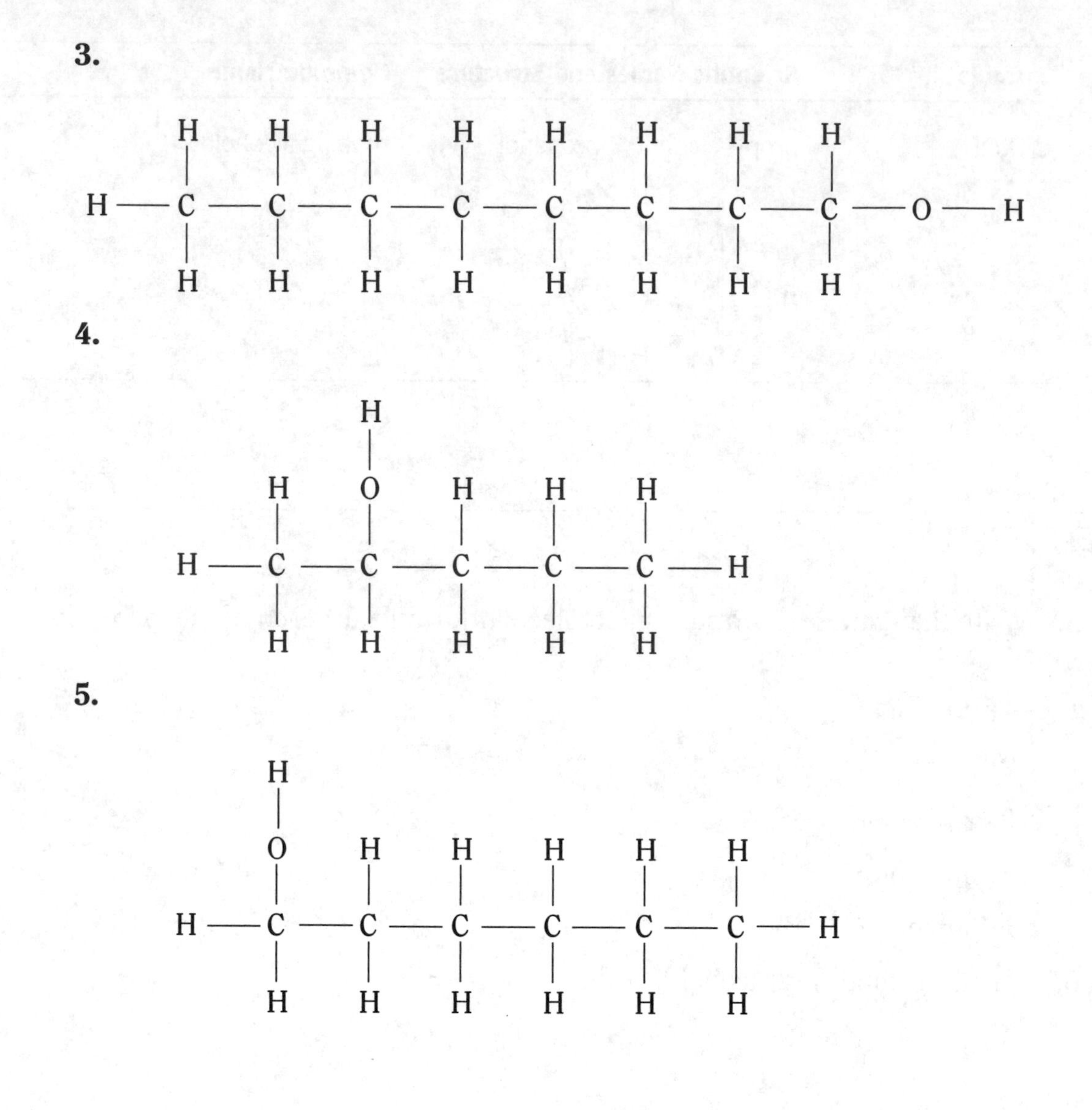

Answers

A.

1.

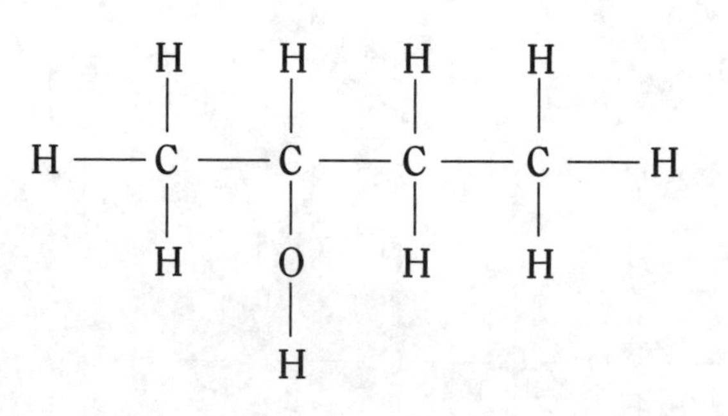

2.

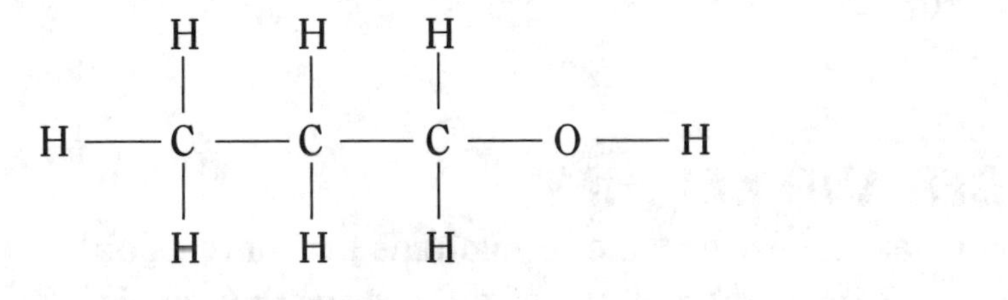

3.

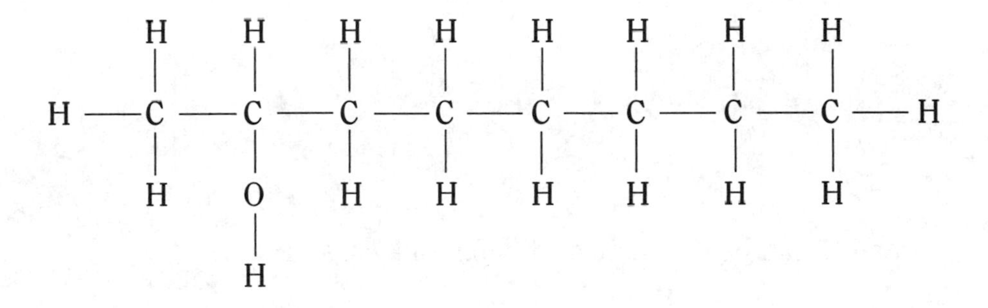

4.

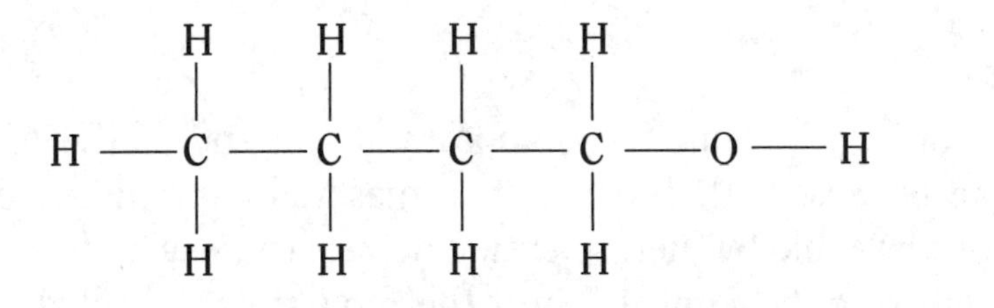

5.

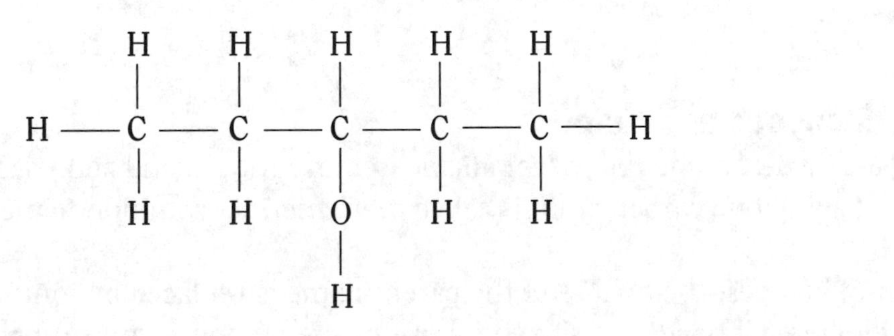

B.

1. 3-heptanol
2. 2-hexanol
3. octanol

4. 2-pentanol
5. hexanol

ALDEHYDES AND KETONES

Aldehydes and ketones are organic compounds that have a double-bonded oxygen atom attached to one of the carbon atoms of the hydrocarbon chain. This C=O combination is called a carbonyl group.

These compounds have the following structural formulas.

$$R-\overset{\overset{\displaystyle O}{\|}}{C}-R'$$

ketone

where R and R' can be the same or different alkyls, and

$$R-\overset{\overset{\displaystyle O}{\|}}{C}-H$$

aldehyde

In aldehydes, the carbonyl group always appears at the end of the carbon chain. Aldehydes familiar to us because of their pleasant aromas include vanilla and cinnamon. If vanilla and cinnamon were sold by their scientific names, there would be a public outcry about the introduction of "chemicals" into the food supply. Embalming fluid, a derivative of formaldehyde, is another aldehyde. It has been suggested that the aldehydes produced along with alcohols in the fermentation process are a contributing factor to hangovers.

Naming Aldehydes and Ketones

In naming aldehydes, name the parent alkane, remove the –e, and add -al. For example, an aldehyde having one carbon atom is called *methanal.* Its common name is formaldehyde.

In naming ketones, the final -e of the parent name is replaced by -one. The location of the double-bonded oxygen is shown by use of the carbon chain numbering system described earlier. For example, in 2-butanone, the carbon chain has four carbon atoms and that double-bonded oxygen atom is attached to the second carbon atom in the chain.

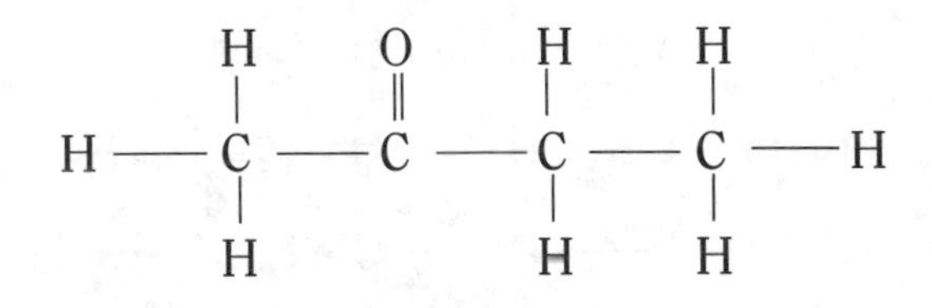

In a consumer sense, propanone is the most familiar ketone; its common name is acetone. The flammability and odor associated with acetone are properties that are common among ketones. They are also generally good solvents—that is often their industrial purpose.

EXERCISES

Complete the following problems dealing with aldehydes and ketones.

1. Draw the structural formula for hexanal.
2. Draw the structural formula for 3-heptanone.
3. Name this structure.

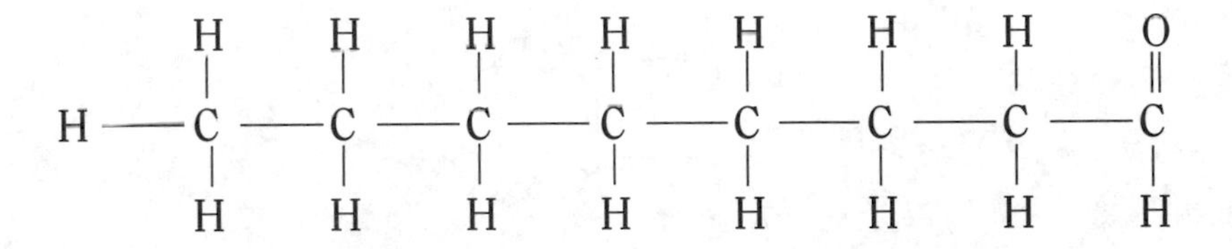

4. Name this structure.

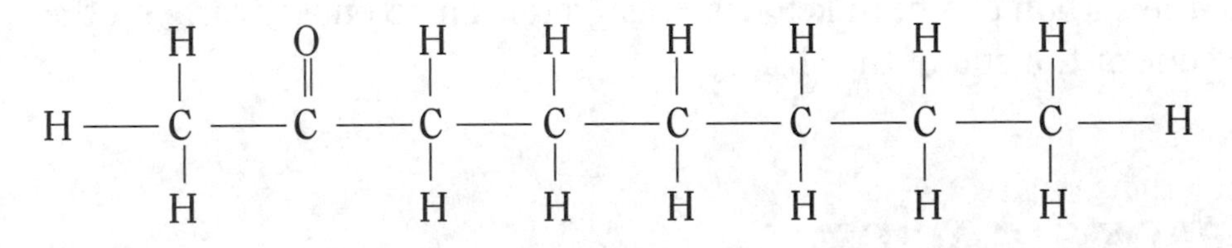

5. Why are there no numbers in aldehyde names?
6. Draw the structure for a carbonyl group.
7. What do aldehydes and ketones have in common?

Answers

1.

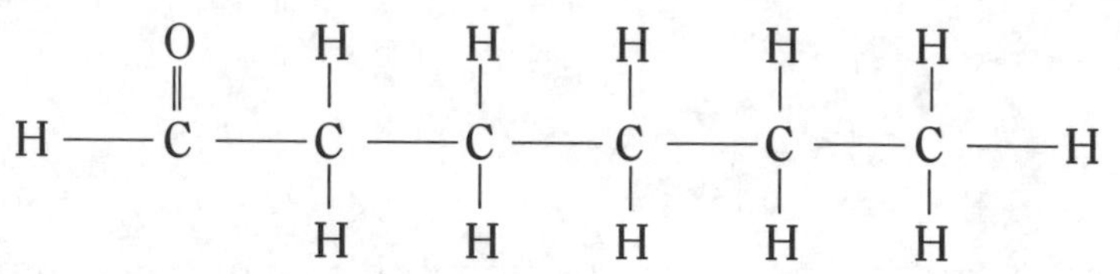

2.

```
      H   H   O   H   H   H   H
      |   |   |   |   |   |   |
  H — C — C — C — C — C — C — C — H
      |   |       |   |   |   |
      H   H       H   H   H   H
```

3. octanal
4. 2-octanone
5. Aldehyde names have no numbers because the aldehyde group is always on the first carbon atom.
6.

```
    O
    ‖
  — C —
```

7. The common denominator linking aldehydes and ketones is the presence of the carbonyl group. In aldehydes this group is on the carbon atom at either end of the carbon chain. In ketones, this group can be on any carbon other than the one at the end of the chain.

CARBOXYLIC ACIDS

Carboxylic acids are characterized by the presence of what is called a carboxyl group, whose molecular formula COOH. Its structural formula is

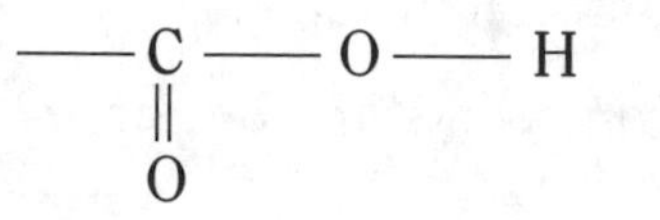

The general formula for a carboxylic acid is RCOOH, where R represents an alkyl group. For example, ethanoic acid has two carbon atoms (eth-), and so its structural formula is

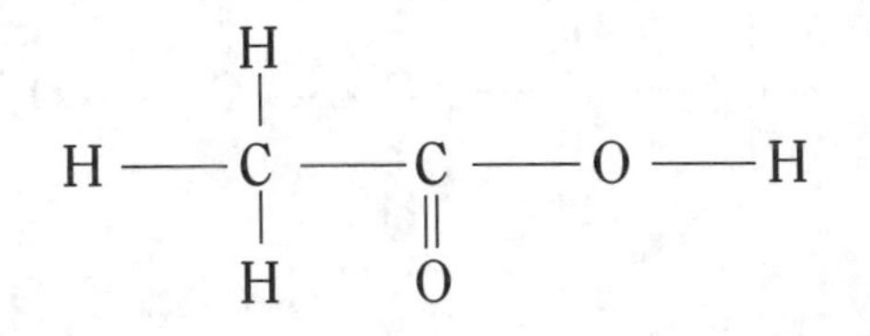

Ethanoic acid, more commonly called acetic acid, is responsible for the familiar odor of vinegar, as vinegar is a five percent solution of ethanoic acid. Methanoic acid has the common name formic acid and the formula HCOOH. Notice that in writing molecular formulas, atoms in the functional group are written together, such as in COOH, to show that an acid is indicated.

Organic acids function as proton donors (H^+) just like the inorganic acids we discussed earlier. The hydrogen donated is the hydrogen that is bonded to the oxygen atom in the carboxyl group:

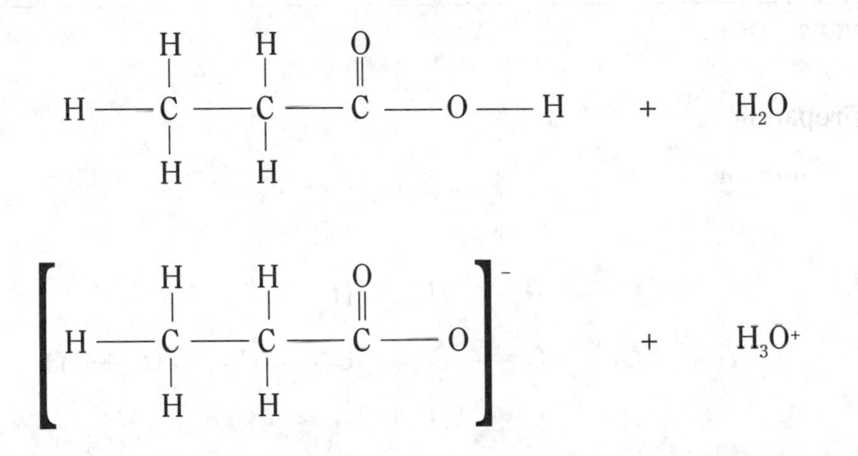

Notice that the hydronium ion (H_3O^+) is formed. Recall that this is the signature ion of the reaction of an acid and water.

EXERCISES

Answer the following questions about carboxylic acids.

1. Name the acid whose structure is shown here.

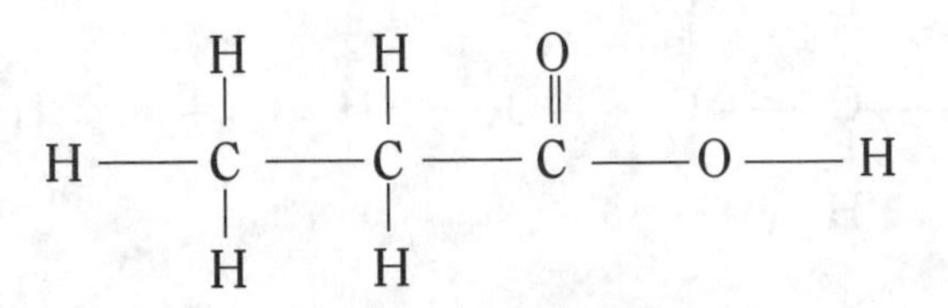

2. Name the acid whose structure is shown here.

```
     H    H    H    O
     |    |    |    ||
H —  C —  C —  C —  C — O — H
     |    |    |
     H    H    H
```

3. Draw the structural formula for pentanoic acid.

4. Draw the structural formula for hexanoic acid.

5. Write a balanced equation for the reaction of water and pentanoic acid.

6. How do an alcohol and an organic acid differ structurally?

Answers

1. propanoic acid

2. butanoic acid

3.

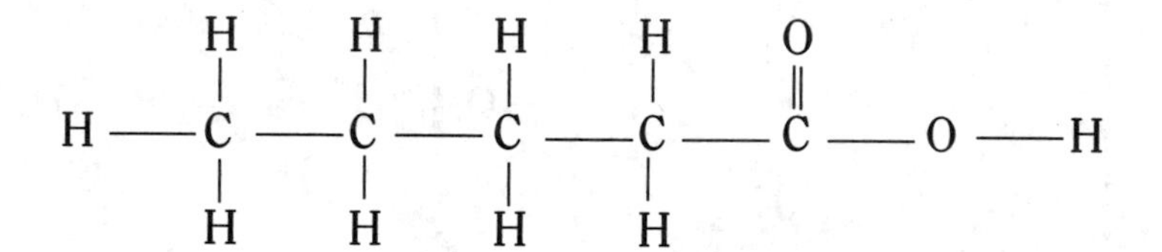

4.

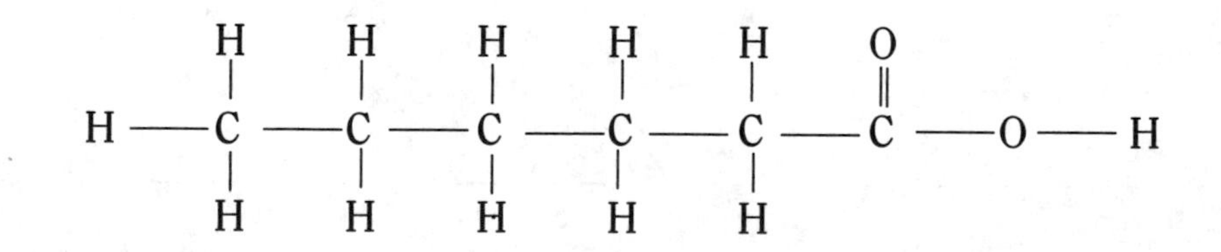

5.

```
     H    H    H    H    O
     |    |    |    |    ||
H —  C —  C —  C —  C —  C — O — H     +     H2O
     |    |    |    |
     H    H    H    H
```

$$\left[H-\underset{H}{\overset{H}{C}}-\underset{H}{\overset{H}{C}}-\underset{H}{\overset{H}{C}}-\underset{H}{\overset{H}{C}}-\overset{O}{\overset{\|}{C}}-O \right]^{-} \quad + \quad H_3O^+$$

6. The functional group OH denotes an alcohol, and the functional group COOH denotes an acid. In addition, the COOH is always be on the terminal carbon of the chain, whereas the OH can be on any of the carbons.

ESTERS

The group of organic compounds designated as esters have the general structural formula

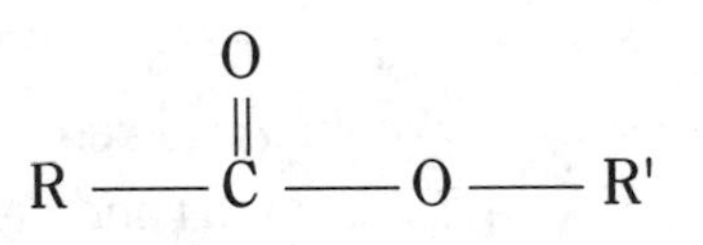

where R and R' can be the same or different alkyl groups. Notice the structural similarity linking esters to organic acids. The R' group takes the place of the acid's ionizable hydrogen. Esters generally have pleasant odors and are primarily responsible for the aromas of fruits.

ETHERS

Ethers are organic compounds having the general structural formula

$$R-O-R'$$

where R and R' can be the same or different alkyl groups. The most familiar ether is diethyl ether, commonly just called ether.

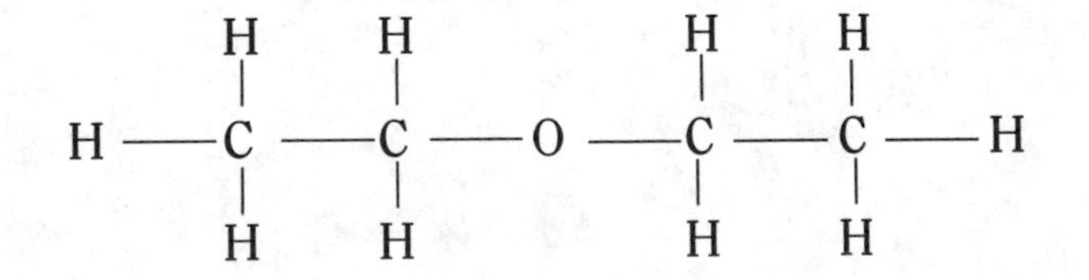

The nature of the bonds between the oxygen and the carbon atoms adjacent to it makes for an unstable molecule. Ethers are highly reactive and require special care and handling.

PETROLEUM

Petroleum is a thick, dark liquid composed mostly of hydrocarbons whose molecules have from 5 to 25 carbon atoms. Natural gas, composed of hydrocarbons whose molecules have 1 to 4 carbon atoms, is often found associated with petroleum deposits. Because petroleum is a mixture of various compounds, in order for it to be used, it must be separated into *fractions* by the refining process. Oil refineries start with the mixture of compounds in the crude oil and by means of distillation and catalysts separate it into more usable fractions. Shorter-chain hydrocarbons boil at a lower temperature than longer-chain hydrocarbons, and so boiling is a way of separating the components of the mixture. Table 13.2 shows the primary uses of the fractions once they are separated.

TABLE 13.2 PRIMARY USES OF PETROLEUM FRACTIONS

Petroleum Fraction	Used for
C_5 to C_{12}	gasoline
C_{10} to C_{18}	jet fuel and kerosene
C_{15} to C_{25}	heating oil and diesel fuel
Greater than C_{25}	asphalt

The petroleum industry and our reliance on it will no doubt continue to primary factors in our economy and daily lives. As events and inventions unfold, organic chemistry will be a major player. The organic chemistry presented in this unit is but the tip of the organic chemistry iceberg, but it is enough to give you a foundation.

UNIT 14

Nuclear Chemistry

Nuclear Symbols and Stability, Fission, Fusion, Half-life, and Nuclear Equations

Up to this point in our discussion of chemistry, reactions have not involved the nuclear part of the atom. Most chemical reactions and behaviors are a function of valence electrons. In general, most nuclei are relatively unreactive. Their interior location coupled with the extreme forces that hold them together make for nuclear stability.

Because nuclear reactions involve protons and neutrons, collectively called *nucleons*, chemists use the notation in Figure 14.1 to show the composition of the nucleus.

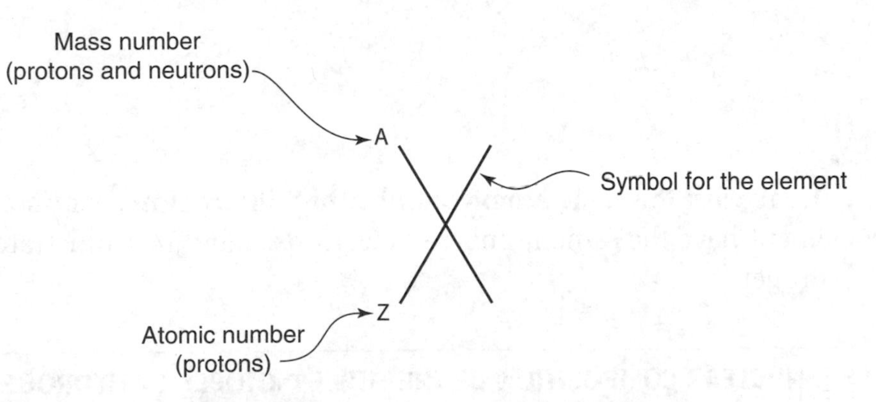

Figure 14.1: Composition of the nucleus

EXERCISES

After examining Figure 14.1, answer the following questions. Use of the periodic table is required. Remember to round mass numbers to the nearest whole number.

1. In terms of *A* and *Z*, how can you find the number of neutrons for a specific nucleus?
2. Given: $^{152}_{63}X$

   How many protons does X have?

   How many neutrons does X have?

   What is the name of the element that X represents?
3. Look up sodium in the periodic table and write its nuclear notation.
4. Write the nuclear notation for bromine.

Answers

1. $A - Z$ = number of neutrons
2. X has 63 protons, and 89 neutrons (subtract the atomic number from the mass number); X is Europium (Eu).
3. $^{23}_{11}Na$
4. $^{80}_{35}Br$

ISOTOPES

***Isotopes* are atoms with the same atomic number but different mass numbers.** All isotopes of an element have the same number of electrons. Table 14.1 illustrates the three isotopes of hydrogen.

TABLE 14.1: NUCLEAR COMPOSITION OF THE THREE ISOTOPES OF HYDROGEN

	Isotope 1	Isotope 2	Isotope 3
Protons	1	1	1
Neutrons	0	1	2
Atomic number	1	1	1
Mass number	1	2	3

Isotopes are designated by their mass numbers. The isotopes in Table 14.1 are called H-1, H-2, and H-3. The various isotopes of any element are naturally occurring, although other isotopes can be produced synthetically. In hydrogen, more than 90 percent of the atoms are the H-1 isotope.

The isotope mixture of each element is the reason that the atomic mass (mass number) is not a whole number. It is actually the weighted average of the isotopes. The mass number listed for hydrogen in the periodic table is 1.008. This value is larger than 1, indicating that there are isotopes heavier than H-1. Because 1.008 is just marginally larger than 1, there is only a small amount of these heavier isotopes.

EXERCISES

Complete the following questions dealing with isotopes.

1. There are three primary isotopes of carbon: C-12, C-13, and C-14. Complete the following table.

Isotope	No. of Protons	No. of Neutrons	Atomic Number	Mass Number
C-12				
C-13				
C-14				

2. Which of the following represent isotopes of the same element?
   (A) 12 protons, 12 neutrons, 12 electrons
   (B) 11 protons, 12 neutrons, 11 electrons
   (C) 12 protons, 13 neutrons, 12 electrons
   (D) 12 protons, 12 neutrons, 13 electrons
3. In the periodic table, sodium is listed with an atomic number of 11 and a mass number of 22.99. Create two possible sodium isotopes, giving their atomic numbers and atomic masses that are consistent with the periodic table values.
4. Which isotope in question 3 is the most abundant in a sample of sodium?

Answers

1.

Isotope	No. of Protons	No. of Neutrons	Atomic Number	Mass Number
C-12	6	6	6	12
C-13	6	7	6	13
C-14	6	8	6	14

2. The correct choices are (A) and (C). These isotopes have the same numbers of protons and different numbers of neutrons. Choice (B) cannot be an isotope because it has a different number of protons than the other choices. Choice (D) is an ion (with a –1 charge) of (A) and (C).

3.

Isotope	No. of Protons	No. of Neutrons	Atomic Number	Mass Number
Isotope X	11	11	11	22
Isotope Y	11	12	11	23

In considering the choices for the above table, both isotopes must have 11 protons, as an isotope with any other number of protons could not be sodium. Because the mass number of sodium is 22.99, one isotope must have a mass number as close to this number (23) as possible, and the other must be less.

4. The most abundant isotope is the one with a mass number of 23 because the atomic mass is so close to this number.

RADIOACTIVITY

A nucleus can undergo a reaction that changes its identity. Some nuclei are unstable, and they become stable by emitting energy and/or particles. **This spontaneous emission from the nucleus of an atom is called radioactivity.**

In considering all the elements and their isotopes, the vast majority are stable and therefore are not radioactive. There is no single rule for predicting which nuclei are radioactive, but there are some general guidelines that provide insight.

Nuclear Stability Guidelines

1. All nuclei with more than 84 protons are likely to be unstable.
2. There are so-called magic numbers associated with the number of nucleons. Nuclei having 2, 8, 20, 50, 82, or 126 protons or neutrons are stable.
3. Nuclei with even numbers of protons and neutrons are more stable than nuclei with odd numbers.

No. of Stable Isotopes	Protons	Neutrons
157	even	even
52	even	odd
50	odd	even
5	odd	odd

4. When the proton number is graphed versus the neutron number for a stable nuclei, a belt of stability emerges (Figure 14.2). Nuclei outside this belt tend to be unstable and therefore radioactive.

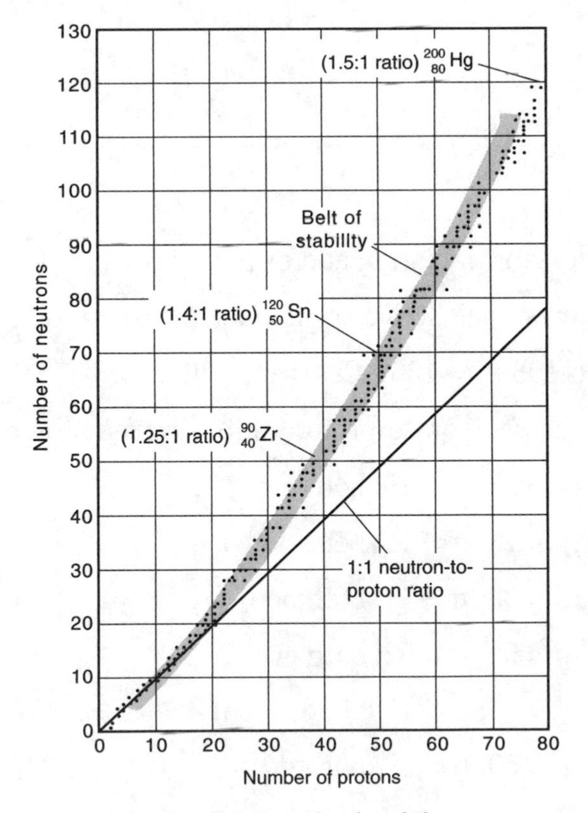

Figure 14.2: Belt of stability

EXERCISES

Answer the following questions dealing with radioactivity.

1. What does a radioactive nucleus emit as it becomes more stable?
2. Would you predict that a sample of francium (Fr) would be radioactive?
3. Which of the following isotopes are stable under the magic number rule?

Cr-52	Br-81
Y-89	Ca-41
S-32	Pb-208

4. Based on an examination of Figure 14.2, is a nucleus having 62 protons and 62 neutrons likely to be stable?
5. Based on the even/odd rule, are any of the following in the most stable category.

Ni-58	Ba-138
Pt-195	Os-190
As-74	Sn-119

Answers

1. Radioactive nuclei emit radiation and/or particles.
2. Francium is likely be radioactive, as its nucleus has more than 84 protons.
3. Y-89 (n = 50); Pb-108 (n = 126); Ca-41 (p = 20)
4. No, this one-to-one neutron-to-proton ratio is outside the belt of stability.
5. Ni-58: yes (p = 28; n = 30; both even)

   Pt-195: no (p = 78; n = 117; one odd)

   As-74: no (p = 33; n = 41; both odd)

   Ba-138: yes (p = 56; n = 82; both even)

   Os-190: yes (p = 76; n = 114; both even)

   Sn-119: no (p = 50; n = 69; one odd)

TYPES OF RADIOACTIVE DECAY

Alpha Decay

Alpha decay is nuclear decomposition such that one of the products of the reaction is an alpha (α) particle, $^{4}_{2}He$. In an example of alpha decay, radium-222 decomposes to form radon-218 plus an alpha particle:

$$^{222}_{88}Ra \rightarrow \ ^{218}_{86}Rn + \ ^{4}_{2}He$$

Inspection of this *nuclear equation* reveals that the total of the mass numbers on the reactant side of the equation equals the total of the mass numbers on the product side. The same is true of the atomic numbers.

Alpha decay improves the stability of radioactive nuclei that lie to the right of the belt of stability. Emission of an alpha particle moves the nucleus diagonally toward the belt of stability because the numbers of both protons and neutrons are decreased by 2. The alpha particle is the least dangerous form of radiation as it has little ability to penetrate tissue.

Beta Decay

Beta decay is characterized by the production of a beta (β) particle, also known as an electron and symbolized as $^{0}_{-1}e$

$$^{234}_{90}Th \rightarrow \ ^{234}_{91}Pa + \ ^{0}_{-1}e$$

Note that in this equation the net effect of beta decay is to change a neutron into a proton. Although an electron is ejected from the nucleus, it was not part of the nuclear composition. The electron called a beta particle comes into being only when the nucleus tries to become stable, as shown here:

$$^{1}_{0}n \rightarrow \ ^{1}_{1}p + \ ^{0}_{-1}e$$

Gamma Decay

A gamma (γ) ray is a high-energy photon of light. The gamma designation of $^{0}_{0}\gamma$ indicates $Z = 0$ and $A = 0$. A nuclide in an excited nuclear energy state can release excess energy by producing gamma rays, very short-wavelength electromagnetic radiation, and high energy. Gamma radiation changes neither the proton nor the neutron count. It often accompanies other radioactive decay because nucleons organizing into more stable configurations need to lose energy.

Positron Emission

A positron is a particle with the same mass as an electron but opposite charge. Symbolically, a positron is $^{0}_{1}e$. A positron does not exist for long. As soon as it encounters an electron, they wipe each other out, forming gamma rays according to the equation

$$^{0}_{1}e \quad + \quad ^{0}_{-1}e \quad \rightarrow \quad 2\,^{0}_{0}\gamma$$

You can think of positron emission as a proton changing to a neutron:

$$^{1}_{1}p \quad \rightarrow \quad ^{1}_{0}n \quad + \quad ^{0}_{-1}e$$

The following equation details the positron emission of sodium-22.

$$^{22}_{11}Na \quad \rightarrow \quad ^{0}_{1}e \qquad ^{22}_{10}Ne$$

Electron Capture

In electron capture, one of the inner-orbital electrons is captured by the nucleus. Gamma rays are always produced along with electron capture.

$$^{201}_{80}Hg \quad + \quad ^{0}_{-1}e \quad \rightarrow \quad ^{201}_{79}Au \quad + \quad ^{0}_{0}\gamma$$

inner-orbital electron

The net result of electron capture is the conversion of a proton to a neutron:

$$^{1}_{1}p \quad + \quad ^{0}_{-1}e \quad \rightarrow \quad ^{1}_{0}n$$

EXERCISES

Using the periodic table and Table 14.2, complete these nuclear equations.

TABLE 14.2: SUBATOMIC PARTICLES

Particle	Symbol
Neutron	${}^{1}_{0}n$
Proton	${}^{1}_{1}p$ or ${}^{1}_{1}H$
Electron	${}^{0}_{-1}e$
Alpha particle	${}^{4}_{2}He$ or ${}^{4}_{2}\alpha$
Beta particle	${}^{0}_{-1}e$ or ${}^{0}_{-1}\beta$
Positron	${}^{0}_{1}e$
Gamma ray	${}^{0}_{0}\gamma$

1. Thorium-230 undergoes alpha decay.

$${}^{230}_{90}Th$$

2. Iodine-131 undergoes beta decay.

$${}^{131}_{53}I$$

3. Nitrogen-13 undergoes positron production.

$${}^{13}_{7}N$$

4. Complete this nuclear equation as an example of electron capture.

$${}^{73}_{33}As$$

5. Complete the following equation and then state the kind of radioactive decay that is occurring.

$${}^{210}_{84}Po \rightarrow {}^{4}_{2}He +$$

6. Write a balanced nuclear equation for the production of a positron by ${}^{11}_{6}C$.
7. Write a balanced nuclear equation for the production of a beta particle by ${}^{214}_{83}Bi$.
8. Write a balanced nuclear equation for the production of an alpha particle by ${}^{237}_{93}Np$.
9. Write a balanced nuclear equation for ${}^{81}_{37}Rb$ undergoing electron capture.

Answers

1. ${}^{230}_{90}Th \rightarrow {}^{4}_{2}He + {}^{226}_{88}Ra$
2. ${}^{131}_{53}I \rightarrow {}^{0}_{-1}e + {}^{131}_{54}Xe$
3. ${}^{13}_{7}N \rightarrow {}^{0}_{1}e + {}^{13}_{6}C$
4. ${}^{73}_{33}As + {}^{0}_{-1}e \rightarrow {}^{73}_{33}Ge$
5. ${}^{210}_{84}Po \rightarrow {}^{206}_{82}Pb + {}^{4}_{2}He$ (alpha decay)
6. ${}^{11}_{6}C \rightarrow {}^{0}_{1}e + {}^{11}_{5}B$
7. ${}^{214}_{83}Bi \rightarrow {}^{0}_{-1}e + {}^{214}_{84}Po$
8. ${}^{237}_{93}Np \rightarrow {}^{4}_{2}He + {}^{233}_{91}Pa$
9. ${}^{81}_{37}Rb + {}^{0}_{-1}e \rightarrow {}^{81}_{36}Kr$

   (orbital electron)

HALF-LIFE

***Half-life* is the time required for half of an amount of radioactive material to decay into a more stable element.** Half-lives for selected isotopes are listed in Table 14.3.

TABLE 14.3: HALF-LIVES OF SOME ISOTOPES

Isotope	Half-life (years)
$^{238}_{92}U$	4.5×10^9
$^{235}_{92}U$	7.1×10^8
$^{40}_{19}K$	1.3×10^9
$^{137}_{55}Cs$	30
$^{131}_{55}I$	0.02

Suppose that a radioactive element has a half-life of 1 day and that you begin with a 2,000-gram sample of it. Table 14.4 shows what happens to this amount over time.

TABLE 14.4: DECAY OF A RADIOACTIVE ELEMENT OVER TIME

After 1 day	1,000 grams remain
After 2 days	500 grams remain
After 3 days	250 grams remain
After 4 days	125 grams remain
After 5 days	62.5 grams remain
After 6 days	31.25 grams remain
And so on	

Notice that with radioactive decay, the radioactive substance never totally disappears. The amount gets smaller and smaller but never becomes zero. This is called *exponential decay*, and a typical graph of it is shown in Figure 14.3.

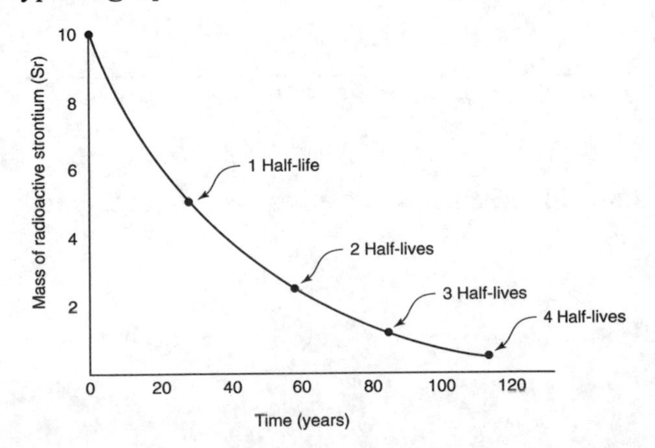

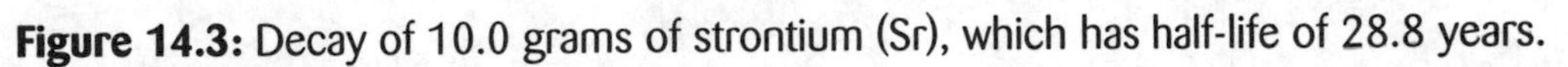
Figure 14.3: Decay of 10.0 grams of strontium (Sr), which has half-life of 28.8 years.

DECAY SERIES

Radioactive decay occurs because of nuclear instability, with the end result of decay being stability. If this stability is not achieved by the first nuclear transformation, then more transformations occur. This set of transformations is called a *decay series,* as shown in Figure 14.4.

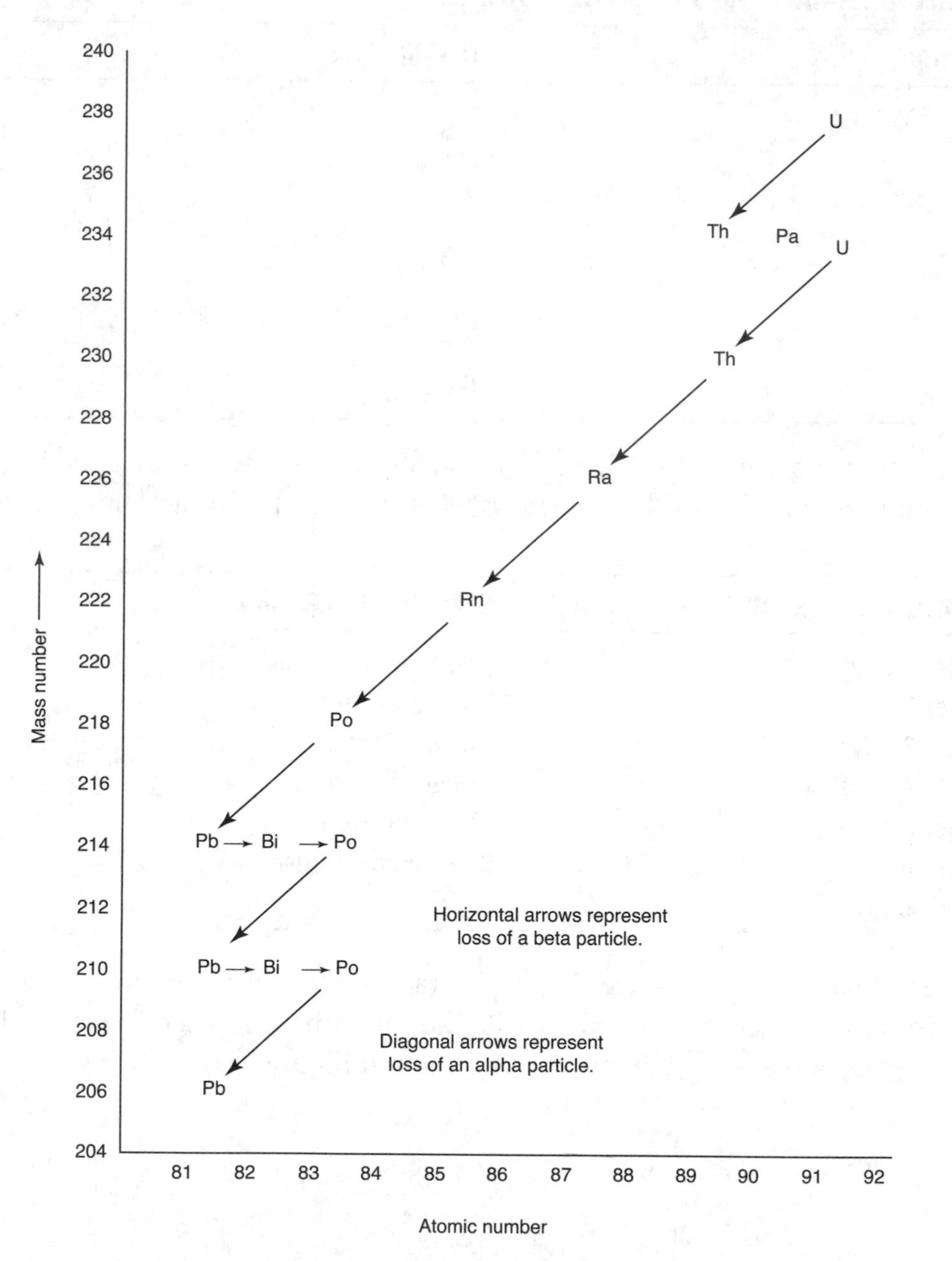

Figure 14.4: Elements formed as uranium decays to lead.

As you examine the decay series in Figure 14.4, note that different elements are formed as uranium decays to become lead. In 1911 Ernest Rutherford was the first to change one element into another.

$$^{14}_{7}N \quad + \quad ^{4}_{2}He \quad \rightarrow \quad ^{17}_{8}O \quad + \quad ^{1}_{1}p$$

Bombarding nitrogen with alpha particles created an isotope of oxygen plus a proton. This success opened the door to the synthesis of hundreds of radioisotopes in the laboratory.

RADIOACTIVE DATING

Radioactive dating is a means of determining the age of a dead plant or animal by comparing the amount of a radioactive isotope in it with the amount of the same radioactive isotope in a living organism. Carbon-14 is often the isotope used for this purpose. Most carbon atoms are the nonradioactive carbon-12 isotope. The radioactive carbon-14 isotope is formed in the atmosphere as a result of random neutron capture, as shown here:

$$^{14}_{7}N \quad + \quad ^{1}_{0}N \quad \rightarrow \quad ^{14}_{6}C \quad + \quad ^{1}_{1}H$$

Carbon-14 becomes part of some of the atmospheric carbon dioxide, which is then incorporated into plants through photosynthesis and finally into animals as they feed on plants. The amount of C-14 in living organisms is fairly constant and therefore provides a baseline for comparison.

Like all radioactive isotopes, C-14 decays at a predictable rate. Its half-life of 5,730 years means that one-half the amount of C-14 normally present in a living organism is present in an organism that has been dead for 5,730 years. By suitable manipulation of the mathematics involved in half-life calculations, the approximate age of the remains of plants and animals can be determined.

NUCLEAR FISSION

Nuclear fission is a nuclear reaction in which nuclei split to form smaller nuclei. The major impact of this splitting process is the enormous energy released. For example, the fission of 1 mole of U-235 (an isotope of uranium) produces 26 million times the energy of the burning of 1 mole of methane. This enormous difference is attributable to liberation of the energy that holds nuclear material together.

The familiar Einstein equation $E = mc^2$ helps explain the enormity of this energy (E). The mass (m) can be a relatively small number, but the value of c is the speed of light (3.0×10^8 meters/second). You can see that by the time the speed of light is squared, to give 9×10^{16}, the value of E is immense, even if the value for the mass is small.

MASS DEFECT

In 1931 scientists discovered that the total mass of the components of a nucleus is greater than the mass of the nucleus itself. This difference is called the *mass defect*. The origin of the mass defect also lies in $E = mc^2$. The components of a nucleus, its protons and neutrons, are bound to each other with an enormous amount of energy. When the nucleus comes apart to form individual protons and neutrons, this binding energy is no longer needed and is converted to mass.

CHAIN REACTIONS

In a *chain reaction*, the fission reaction keeps going because one of the products of the reaction is used to start a new reaction. Notice in the fission reaction below that one neutron (${}^{1}_{0}n$) is on the reactant side and three neutrons ($3\,{}^{1}_{0}n$) are on the product side of the equation. These three neutrons can go on to initiate other reactions.

$${}^{1}_{0}n \quad + \quad {}^{232}_{92}U \quad \rightarrow \quad 3\,{}^{1}_{0}n \quad + \quad {}^{142}_{56}Ba \; + \; {}^{91}_{36}Kr$$

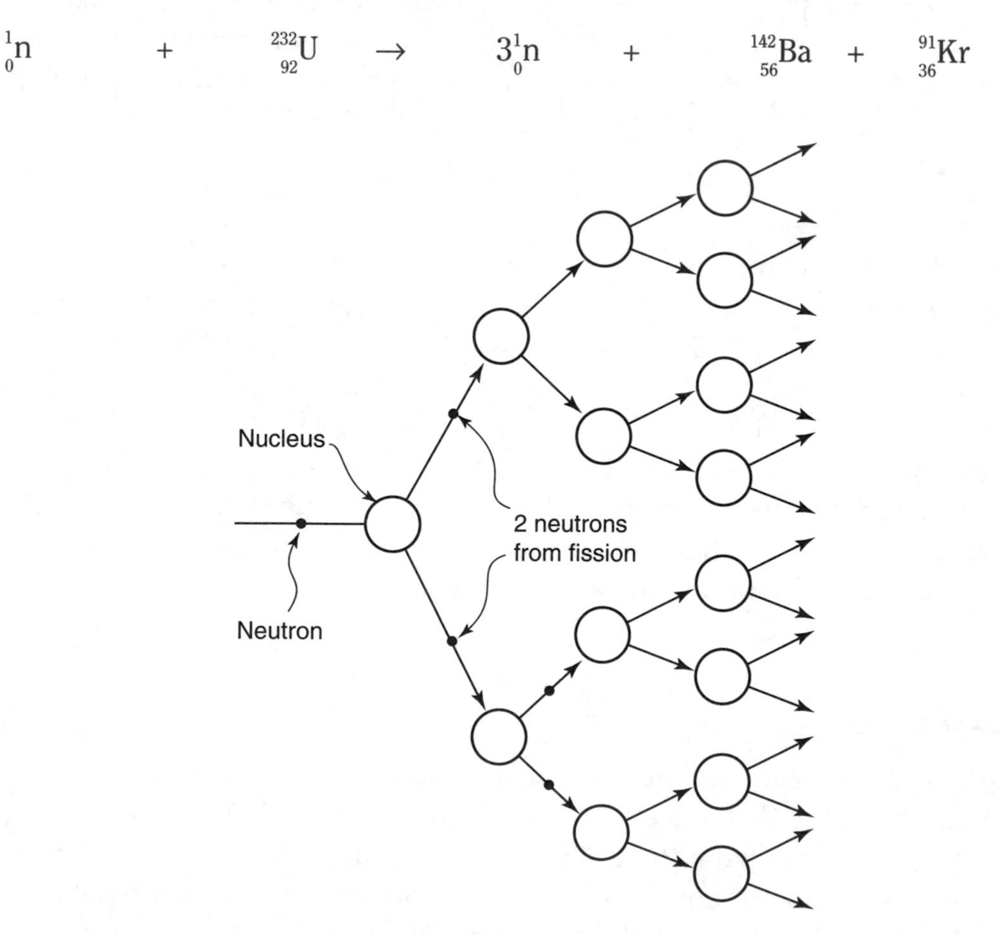

Figure 14.5: Chain reaction

CRITICAL MASS

In the chain reaction shown in Figure 14.5, if there is so little uranium that the released neutrons escape before they have a chance to cause a fission reaction, the reaction stops. **The critical mass is the minimum mass of fissionable material needed in order for the reaction to continue.** The critical mass concept is the key to the design of a fission-type nuclear weapon. In such a weapon, two smaller-than-critical masses are present but are separated. When these subcritical masses are suddenly combined, the rapidly escalating fission reactions produce an explosion of incredible intensity.

NUCLEAR REACTORS

The enormous amounts of energy produced from nuclear reactions makes them a logical choice for energy production. Essentially, the generation of electricity by nuclear means occurs as follows.

- The controlled nuclear reaction produces heat.
- This heat is used to boil water.
- The steam thus created is used to drive a turbine.
- The turbine generates electricity.

Nuclear reactors obviously bypass the use of fossil fuels such as coal, oil, and natural gas in the production of electricity to satisfy the growing energy demands of the United States. Fossil fuels have their own drawbacks, including pollution and an ultimate limit on their availability. The most feared concern associated with nuclear power plants is a meltdown. A failure of the cooling system can result in temperatures high enough to melt the reactor core. Despite assurances by nuclear designers, the accidents at Three Mile Island (Pennsylvania in 1979) and Chernobyl (Russia in 1986) have certainly been frightening. The disposal of spent nuclear materials is the other primary concern associated with nuclear power plants.

NUCLEAR MEDICINE

Nuclear medicine is a subspecialty in the field of radiology. The images developed in nuclear medicine are based on the detection of energy emitted by a radioactive substance (radiopharmaceutical agent) given to a patient. These substances are administered to the patient and then followed as they move through the body and are utilized by the body's chemistry. For example, a radioactive isotope of iodine is used as a diagnostic thyroid test. The thyroid gland is the body's only important user of iodine, and so the efficiency of this gland can be measured by measuring its radioactivity.

NUCLEAR FUSION

Nuclear fusion is the combining of two light nuclei to form a heavier nucleus:

$$4\,{}^{1}_{1}H \rightarrow {}^{4}_{2}He + 2\,{}^{0}_{1}e$$

As in nuclear fission, incredible amounts of energy are produced from relatively small amounts of materials. From a raw materials point of view, nuclear fusion is ideal. Hydrogen and helium are readily available fusion reactants. Fusion reactions power the sun, which helps to explain the difficulty in creating the conditions that would allow fusion to take place. The astronomically high temperatures required present technical problems that may never be resolved.

Whatever the future of the various forms and uses of nuclear energy, it is certain that much of the future will have a nuclear component.

UNIT 15

Chemical Calculations

Calculator Use, Algebra for Chemistry, Significant Figures, and the Use of Mathematical Formulas

In this unit you will find explanations, examples, and practice dealing with the calculations encountered in the chemistry discussed in this book. The types of calculations included here involve conversion factors, metric use, algebraic manipulations, scientific notation, and significant figures. This unit can be used by itself or be incorporated for assistance with individual units. Unless otherwise noted, all answers are rounded to the hundredth place. The calculator used here is a Casio FX-260. Any calculator that has a log (logarithm) key and an exp (exponent) key is sufficient for these chemical calculations.

CONVERSION FACTORS

The conversion factor problem-solving technique has been used throughout this book, especially in the units on moles and stoichiometry. These problem solutions are generally in a format like this:

$$A \times \frac{B}{C} \times \frac{D}{E} = F$$

The plan for entering these values into the calculator is

$$A \times B \div C \times D \div E =$$

The general idea is that every numerator value is multiplied and every denominator value is divided. For example:

$$25 \times \frac{2}{3} \times \frac{13}{17} =$$
$$25 \times 2 \div 3 \times 13 \div 17 = 12.75$$

Note that it is not necessary to press the equal key until the end.

EXERCISES

1. $300 \times \frac{52}{18} \times \frac{5}{17} =$

2. $50{,}000 \times \frac{350}{400} \times \frac{273}{325} \times \frac{200}{425} =$

3. $94 \times \frac{7}{12} \times \frac{13}{23} \times 413 =$

Answers

1. 254.90
2. 17,294.12
3. 12,800.01

METRIC CALCULATIONS

The metric system is the system of choice for chemistry, as well as for other physical sciences such as physics and astronomy. You may not feel comfortable with this system, but two aspects of its usage should relieve your concerns. First, the most troublesome mathematical manipulation, changing back and forth between the metric system and the English system (pounds, miles, gallons, and so on), is rarely required and is not included here. Second, less frequently used metric units, such as decimeters, need not be part of what you need to learn.

The heart of the metric system is its use of prefixes. Many of the prefixes are ones that you already know. Table 15.1 lists all of the ones that you need to know.

TABLE 15.1: METRIC PREFIXES

Prefix	Meaning	Example	
nano-	billionth	nanogram	0.000000001 gram
micro-	millionth	microgram	0.000001 gram
milli-	thousandth	milliliter	0.001 liter
centi-	hundredth	centimeter	0.01 meter
kilo-	thousand	kilogram	1,000 gm

Metric units are often abbreviated, as shown in Table 15.2.

TABLE 15.2: METRIC ABBREVIATIONS

Metric Unit	Abbreviation
nanogram	ng
microgram	g
milligram	mg
gram	g
kilogram	kg
milliliter	ml
liter	l
meter	m
centimeter	cm
millimeter	mm

EXERCISES

Use the prefixes in Table 15.1 to answer these questions.

1. How many meters are in a kilometer?
2. A millimeter is what fraction of a meter?
3. How many centimeters are there in a meter?
4. How many nanograms are there in 1 gram?
5. Which is larger, a nanogram or a microgram? By how much?
6. How many milliliters are in 1 liter?

Answers

1. 1,000
2. 0.001 or 1/1000
3. 100
4. 1,000,000,000
5. A microgram—a thousand times more, which is the difference between 0.000001 and 0.000000001.
6. 1,000

The conversion factor approach is quite useful with metric problems. (If you have not yet worked your way through the conversion factor presentation in Unit 4, you might want to do so now.) The metric conversion factors are ones that you can make yourself using the prefixes in Table 15.1. Examples include

$$\frac{1\text{ meter}}{100\text{ cm}} \quad \frac{1\text{ centimeter}}{10\text{ mm}} \quad \frac{1{,}000\text{ grams}}{1\text{ kg}} \quad \frac{1\text{ kilogram}}{1{,}000\text{ g}} \quad \frac{1\text{ liter}}{1{,}000\text{ ml}} \quad \frac{1{,}000\text{ ml}}{1\text{ liter}}$$

Here is an example of a metric problem: How many kilometers are equal to 15 cm?

$$15\text{ cm} \times \frac{1\text{ m}}{100\text{ cm}} \times \frac{1\text{ km}}{1{,}000\text{ m}} = 0.00015\text{ or }1.5 \times 10^{-4}\text{ km}$$

Recall the following procedure from the discussion of conversion factors in Unit 4.

- The solution begins with the number and the unit of measure from the problem itself.
- The next conversion factor has as its denominator units the same units that the problem began with.
- Units are canceled when one appears in a numerator and one appears in a denominator.
- Conversion factors are continued until all the units have canceled out except for the units desired for the answer.
- Review the solution setup above to see that these steps have been followed.

Examples

1. Using the conversion factor plan just described, determine how many grams of a prescription drug are equal to 5 mg.

$$5\ \cancel{mg} \times \frac{1\ g}{1{,}000\ \cancel{mg}} = 0.005\ g$$

2. How many kilometers are equal to 350 millimeters?

$$350\ \cancel{mm} \times \frac{1\ \cancel{m}}{1{,}000\ \cancel{mm}} \times \frac{1\ km}{1{,}000\ \cancel{m}} = 0.00035 \text{ or } 3.5 \times 10^{-4}$$

EXERCISES

Solve the following metric problems.

1. 55 grams = ? kilograms
2. 50 centimeters = ? meters
3. 0.3 liter = ? milliliters
4. 3.7 meters = ? millimeters
5. 0.08 kilogram = ? micrograms
6. 4500 milliliters = ? liters

Answers

1. $55\ \cancel{grams} \times \frac{1\ kilogram}{1{,}000\ \cancel{grams}} = 0.055\ kilogram$

2. $50\ \cancel{cm} \times \frac{1\ meter}{100\ \cancel{cm}} = 0.5\ meter$

3. $0.3\ \cancel{liter} \times \frac{1{,}000\ ml}{1\ \cancel{liter}} = 300\ ml$

4. $3.7 \cancel{\text{meters}} \times \frac{1{,}000 \text{ mm}}{1 \cancel{\text{meter}}} = 3{,}700 \text{ mm}$

5. $008 \cancel{\text{kg}} \times \frac{1{,}000 \cancel{\text{grams}}}{1 \cancel{\text{kg}}} \times \frac{1{,}000 \text{ mg}}{1 \cancel{\text{g}}} = 80{,}000 \text{ grams}$

6. $4{,}500 \cancel{\text{ml}} \times \frac{1 \text{ liter}}{1{,}000 \cancel{\text{ml}}} = 4.5 \text{ liters}$

ALGEBRA MANIPULATIONS

The algebra involved in the chemistry problems you encounter in this book falls into two categories.

Category One: *AB* = *CD*

In this kind of equation, the unknown can be in any one of the four letter positions.

Examples

1. $(15)x = (25)(3)$

   To obtain x by itself, divide both sides of the equation by 15.

   $$\frac{(15)x}{15} = \frac{(25)(3)}{15}$$
   $$x = 5$$

2. $(35)(7) = (27)x$

   $$\frac{(35)(7)}{27} = \frac{(27)x}{27}$$
   $$9.07 = x$$

EXERCISES

Solve the following problems for x.

1. $200x = (175)(5.5)$
2. $(3.7)(1.5) = x(26.1)$
3. $x(207) = 18.1(198)$
4. $273(27.9) = 318x$

Answers

1. 4.81
2. 0.21
3. 17.31
4. 23.95

Category Two: $\frac{A}{B} = \frac{C}{D}$

The unknown can be in any one of the four letter positions in the equation. In solving this kind of equation, the first step is to cross-multiply:

$$AD = BC \quad \text{or} \quad BC = AD$$

From this point on, the method of solving is the same as for problems in the previous category. For example, when solving for A, both sides of the equation are divided by D.

Example

$$\frac{30.2}{273} = \frac{29.8}{x}$$

$$30.2x = (273)(29.8)$$

$$x = \frac{(273)(29.8)}{30.2}$$

$$= 269.38$$

EXERCISES

Solve these problems for x.

1. $\frac{x}{15} = \frac{50}{40}$
2. $\frac{8}{50} = \frac{x}{75}$

3. $\frac{300}{x} = \frac{200}{12}$

4. $\frac{70}{50} = \frac{95}{x}$

Answers

1. 18.75
2. 12
3. 18
4. 67.86

SCIENTIFIC NOTATION

Scientific notation is a method of writing numbers in a specific format that is best explained by presenting an example. Suppose you want to write the number 2,376 in scientific notation. The first step is to rewrite the number putting the decimal immediately after the first digit, as in 2.376. Obviously 2,367 is not the same size number as 2.376, and so the 2.376 has to be multiplied by a power of 10. If 2.376 is multiplied by 1,000, we get the original number 2,376. Because 1,000 can be written as 10^3, the 2,376 can be written in scientific notation as 2.376×10^3. This format is called scientific notation. The goal is always for the number written in scientific notation to have the same value as the original number.

The following examples show how to convert ordinary numbers to scientific notation.

$$54 = 5.4 \times 10^1$$
$$70{,}789 = 7.0789 \times 10^4$$
$$0.0898 = 8.98 \times 10^{-2}$$

Note that a negative exponent moves the decimal to the left and a positive exponent moves the decimal to the right.

The following examples show how to convert numbers from scientific notation to ordinary numbers.

$$2.44 \times 10^5 = 244{,}000$$
$$7.053 \times 10^{-2} = 0.07053$$
$$6.8 \times 10^1 = 68$$

EXERCISES

A. Write the following numbers in scientific notation.

1. 45,600 =

2. 98.22 =

3. 0.00354 =

4. 0.9234 =

5. 10,000 =

6. 0.001 =

B. Convert these numbers from scientific notation to ordinary numbers.

1. 3.006×10^{-5} =

2. 4.1×10^{6} =

3. 9.0112×10^{-3} =

4. 6.003×10^{2} =

5. 1×10^{-5} =

6. 1×10^{2} =

Answers

A.

1. 4.5600×10^{4}

2. 9.822×10^{1}

3. 3.54×10^{-3}

4. 9.234×10^{-1}

5. 1×10^{4}

6. 1×10^{-3}

Answer

1. 0.00003006
2. 4100000
3. 0.0090112
4. 600.3
5. 0.00001
6. 100

Entering Scientific Notation into a Calculator

The next step is entering a number written in scientific notation into a calculator. Locate the EXP key. For practice, enter the frequently used chemistry number 6.02×10^{23} by performing the following steps.

- 6.02
- EXP
- 23

It is important to note that you did not enter the 10, nor did you enter the multiplication sign. If you experiment and enter the number into your calculator just as it is written, that is, 6.02 times 10 exponent 23, you will get an answer that is 10 times too big, or 6.02×10^{24}. It is important that you realize that the calculator uses a short cut by showing 6.02^{23} when it really means 6.02×10^{23}. Some tests allow you to use this shortcut, but it is always safe to use the true representation, which is the 10 with its proper exponent.

Scientific Notation in Problem Solving

The universally accepted practice is to wait until the final answer is displayed on the calculator screen before rounding.

EXERCISES

Solve the following problems, rounding the answer to the hundredth place and expressing the answer in scientific notation.

1. $\dfrac{(5.5\times10^5)(2.1\times10^3)}{7.8\times10^3}$

2. $\dfrac{(6.02\times10^{23})(23)}{10^3}$

3. $\dfrac{5.0\times10^3}{(6.02\times10^{23})(10)}$ (*Hint:* Divide by the denominator 10.)

4. $\dfrac{(2.0\times10^5)(5\times10^{-1})}{2.5\times10^7}$

5. $\dfrac{(6.02\times10^{23})(5.5\times10^4)}{(6.5\times10^2)(2.1\times10^5)}$

Answers

1. 1.48×10^5
2. 1.38×10^{22}
3. 8.31×10^{-22}
4. 4×10^{-3}
5. 2.43×10^{20}

SIGNIFICANT FIGURES

The idea of significant figures is based on what is called uncertainty. If you step on a bathroom scale and see your weight displayed as 149 pounds, you might really weigh 148.6 pounds and the 149 is a rounded number. You could just as easily weigh 149.4 pounds, with the scale rounding this value to 149. This bathroom scale measures weight to the first uncertain figure. A *significant figure* is defined as all the digits that you are sure of plus the first uncertain digit.

As you look at Figure 15.1, what value would you give to the position marked by the arrow?

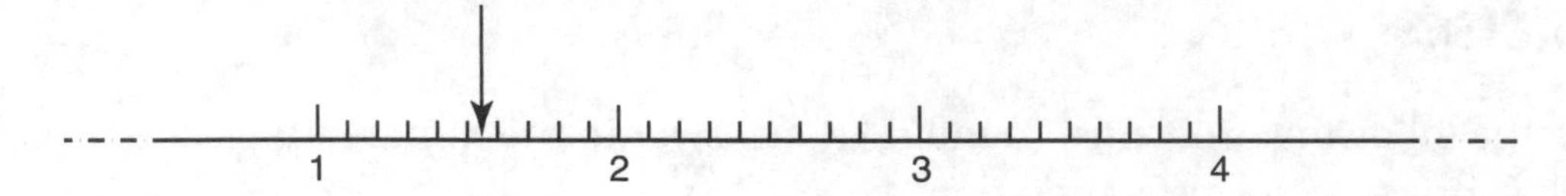

We can all agree that the arrow is on 1.5 something, but 1.54, 1.53, and 1.52 are all possible choices. The 1 and the 5 are certain, but the next place is the first uncertain figure. Therefore, it is correct to read the ruler to two decimal places because the hundredth place is the first one that is uncertain. If someone records the value of the arrow's position as 1.545, this value will be an incorrect statement of the measuring ability of the ruler.

Suppose you use this ruler to measure a box with the dimensions shown in Figure 15.2.

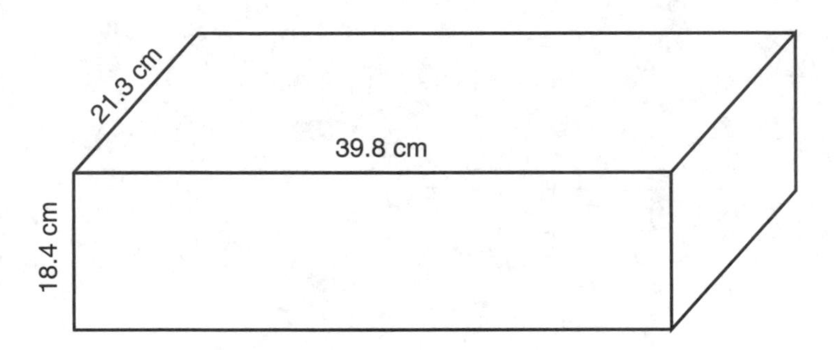

Calculate the volume of the box:

$$\begin{aligned} V &= \text{(length)(width)(height)} \\ &= (39.8\text{ cm})(21.31\text{ cm})(18.4\text{ cm}) \\ &= 15{,}598.416\text{ cm}^3 \end{aligned}$$

But can we promise so much accuracy from a ruler that can measure with certainty only one place past the decimal? It turns out that the best we can say is that the volume of this cube is 16,600 cm^3. The reason for rounding to 16,600 is explained in the next section.

Counting Significant Figures

It is important that you know how to count the number of significant figures in any measurement, as that shows you how to report the results of using a particular measuring device. The rules for determining the number of significant figures in a measurement are

- Nonzero integers are always significant.
- Leading zeros that come before all nonzero digits are never significant. For example, in the number 0.0099, none of the zeros is significant and both 9s are significant; so the number has two significant figures.

- Captive zeros are trapped between two nonzero digits, as in 5.006. Captive zeros are always significant. In 5.006 there are four significant figures.
- Trailing zeros are at the right end of a number, as in 23.9000. They are significant only if the number has a decimal somewhere in it. In 23.9000, all the zeros are significant, and so the number itself has six significant figures. However, if a number like 35,000 has no decimal, its zeros are not significant, and 35,000 has only two significant figures.

Note the following examples of the number of significant figures.

34,981 has five significant figures; all the digits are nonzero.

3.0898 has five significant figures; the zero is a captive zero.

24,000 has two significant figures; the zeros are trailing zeroes in a number without a decimal.

24,000 has five significant figures; the zeros are trailing zeroes in a number with a decimal.

0.056 has two significant figures; the zeros are leading.

EXERCISES

How many significant figures are in each of the following numbers?

1. 87,009
2. 0.0987
3. 88.00
4. 20040
5. 14.10

Answers

1. five significant figures; captive zeros.
2. three significant figures; leading zeros are not significant.

3. four significant figures; trailing zeros are significant because there is a decimal point.
4. four significant figures; captive zeros are significant, but the trailing zero is not.
5. four significant figures; trailing zero is significant because of the decimal point.

Now that you have learned to count significant figures, the next (and final) step is to learn the rule for rounding answers to obtain the correct number of significant figures. The rule for multiplication and division is that the answer can have only the same number of significant figures as the least significant data used to produce that answer. If, for example, you multiply three measurements together and one of them has only two significant figures, the answer must be rounded to two significant figures. This is what happened for the volume of the box in the example above. Each piece of data had only three significant figures, and so the answer had only three significant figures.

Example

$$\frac{(2.010)(45)}{3.005} = 30.09983361 = 30$$

The piece of data with the fewest number of significant figures is the 45, with only two significant figures. Therefore, the answer must be rounded to only two significant figures. The decimal must be placed in the final answer because without it there would be only one significant figure.

EXERCISES

Perform the following calculations, rounding the answer to the correct number of significant figures.

1. $(23.090)(15) =$
2. $\frac{(273)(3.02)}{0.25} =$
3. $\frac{(50,000)(3.008)}{24.7} =$
4. $(0.0015)(3.006) =$

Answers

1.	346.35	=	350.	Round to two significant figures because of the 15.
2.	3,297.84	=	3,300.	Round to two significant figures because of the 0.25.
3.	6089.068826	=	6,000.	Round to one significant figure because of the 50,000.
4.	0.004509	=	0.0045.	Round to two significant figures because of the 0.0015.

There is another rule for addition and subtraction, but its rarity in chemical use makes it not worth our time here.

Index